Vincent Stephen

Wrinkles in Electric Lighting

Vincent Stephen

Wrinkles in Electric Lighting

ISBN/EAN: 9783337249489

Printed in Europe, USA, Canada, Australia, Japan

Cover: Foto ©berggeist007 / pixelio.de

More available books at **www.hansebooks.com**

WRINKLES

IN

ELECTRIC LIGHTING.

WRINKLES

IN

ELECTRIC LIGHTING.

BY

VINCENT STEPHEN.

E. & F. N. SPON, 125, STRAND, LONDON.
NEW YORK: 12, CORTLANDT STREET.
1888.

INTRODUCTION.

In the following pages it is my intention to give engineers on board ship, who may be put in charge of electric lighting machinery without having any electrical knowledge, some idea of the manner in which electricity is produced by mechanical means; how it is converted into light; what precautions must be used to keep the plant in order, and what to do in the event of difficulties arising. I do not therefore aim at producing a literary work, but shall try and explain everything in the plainest language possible.

CONTENTS.

THE ELECTRIC CURRENT, AND ITS PRODUCTION BY CHEMICAL MEANS.

PAGE

Production of electric current in chemical battery—Current very weak—Current compared to circulation of the blood—Strength and volume of current—Pressure not sufficient without volume—Action of current is instantaneous—Resistance to the passage of the current—Copper the usual metal for conductors—Heat produced by current when wire is too small 1

PRODUCTION OF ELECTRIC CURRENTS BY MECHANICAL MEANS.

Current produced by mechanical means — Alternating current—Magneto-electric machines—Shock produced by interruption of current—The current must be commutated—Description of commutator—Current, though alternating in the dynamo, is continuous in the circuit—Continuous current used for electro-plating 5

Dynamo-Electric Machines.

Current will magnetise an iron or steel bar—Permanent magnet — Electro-magnet — Where the magneto and dynamo machines differ—Armature of so-called continuous-current dynamo—Type of commutator—Commutator brushes—Current continuous in the circuit—Alternating-current dynamos—Current not commutated

WRINKLES
IN
ELECTRIC LIGHTING.

THE ELECTRIC CURRENT, AND ITS PRODUCTION BY CHEMICAL MEANS.

IT will first be necessary to explain how electric currents are produced by means of chemicals. In a jar A, Fig. 1, are placed two plates B and C, one zinc, and the other copper, each having connected to it at the top a copper wire of any convenient length. The plates are kept in position by means of pieces of wood, and the jar is about half filled with a solution of salt and water, or sulphuric acid and water; if then the two wires are joined, a current of electricity at once flows through them, however long they may be. The current produced in this manner is very weak, and does not even keep what strength it has for any length of time, but rapidly gets weaker until quite imperceptible. The

Production of electric current in chemical battery.

Current very weak.

current is, however, continuous; that is, it flows steadily in the one direction through the wire, and may be used for ringing bells, or for other purposes where a feeble current only is required to do intermittent work. The wire E in connection with the copper plate is called the positive lead, and the other the negative, and the current is said to flow from the copper plate, through the wire E through the circuit to D, and thence to the zinc plate, and through the liquid to the copper plate. The current has often been compared to water flowing through a pipe, but I think it can be better compared to the blood in the human body, which through the action of the heart is continually forced through the arteries and veins in one steady stream. There is, however, this difference, that there is no actual progression of matter in the electric current, it being like a ripple on water, which moves from end to end of a lake without the water itself being moved across. Now that I have given you an idea of how the current acts, I must try and explain how different degrees of strength and volume are obtained. In the first place, let us consider what constitute strength and volume in an electric current, or at least try and get a general notion about them. For this purpose I shall compare the electric current to water being forced through a pipe; and the strength of the electric current, or electromotive force, written for short E.M.F., will be like the pressure of

Margin notes: Current compared to circulation of the blood. Strength and volume of current.

water at any part of the pipe. Two pipes may carry different quantities of water, and yet the pressure may be the same in each; in one a gallon of water may pass a given point in the same time that a pint passes the same point in the other, and yet in each case the different quantities may pass that point at the same speed. Thus in electricity, two currents may be of different volume or quantity, measured in ampères, and yet be of the same E.M.F. measured in volts; or they may be of different E.M.F., or pressure, or intensity, and yet be of the same volume. If any work is to be done by the water forced through a pipe, such as turning a turbine, it is evident that pressure of itself is not sufficient, seeing that a stream an inch in diameter may be at the same pressure as another a foot in diameter. So with the electric current, if work is to be done, such as driving a motor or lighting a lamp, it is not sufficient to have a certain E.M.F.; there must be quantity or volume in proportion to the amount of work, so that if it takes a given quantity to work one lamp, it will take twice that quantity to work two lamps of the same kind. It must not be inferred from this, that if one lamp requires a certain E.M.F., that two lamps will require it to be doubled, as such is not the case, except under certain conditions which I will explain later on. *Pressure not sufficient without volume.*

The action of electricity is practically instantaneous in any length of wire, so that if the current *Action of current is instantaneous.*

is used to ring two bells a mile apart, but connected by wires, they will commence to ring simultaneously. I have so far not said anything about resistance to the passage of the current through the wires. I shall therefore refer again to our comparison of the current to water forced through a pipe, and you will agree that a certain sized pipe will only convey a certain amount of water in a given time. If a larger quantity is to be conveyed in the same time, a greater pressure must be applied, or a larger pipe must be used.

It is evident that increasing the size of the pipe will get over the difficulty more readily than increasing the pressure of the water. The pipes themselves offer a certain resistance to the passage of the water through them, in the shape of friction; so that if an effect is to be produced at a distance, rather more pressure is required than if it is done close at hand, so as to make up for the loss sustained by friction.

Resistance to the passage of the current. Much the same may be said of the electric current; a certain sized wire will only carry a certain current, and if more current is required, a thicker wire must be used to convey it, or it must be of a greater E.M.F. It is usually more convenient to increase the thickness of the wire than to increase the E.M.F. of the current. The wire offers a certain resistance to the passage of the current through it, which may be compared to friction, and this resistance varies according to the metal of which it is composed. Copper is the metal in ordinary use for wires for electric lighting purposes, and the purer it is the better will it convey the current. Iron is used for telegraph wires on account of cheapness, the current used being so small that this metal conveys it readily enough; if copper were used, the wires will only require to be about one-third the diameter of the iron ones. The following are the respective values for electrical conductivity of various metals when pure, taking silver as a standard:—Silver 100, copper 99·9, gold 80, zinc 29, brass 22, iron 16·8, tin 13·1, lead 8·3, mercury 1·6.

Copper the usual metal for conductors.

If a wire is made to convey a current which is too large for its electrical capacity, it will get heated, which decreases its conductivity, with the result that the heat increases until finally the wire fuses. I shall have more to say about this when speaking of electric lighting.

Heat produced by current when wire is too small.

Production of Electric Currents by Mechanical Means.

Magneto-electric Machines.

I have shown how the electric current is produced by the action of chemical or primary batteries, and how this current will flow through suitable conductors. I shall now explain how mechanical power may be converted into electricity. It has been found that if a wire, preferably of copper, of which the ends are joined together, is moved past a magnet a current is induced in the wire, flowing in one direction while the wire is approaching the magnet, and in the opposite direction while it is receding from it. This is then not a continuous current like we obtained from the chemical battery, but an alternating one, and you will see later on how it can be made to produce similar effects. The oftener the wire passes the magnet the more electricity is generated, so that if we make a coil of the wire and move a large number of parts of wire past at one time, the effects on each part are accumulated; and if instead of having one magnet to pass before, we have several, the effects will be doubled or trebled, &c., in proportion to the number. If, again, the coil is moved at an increased speed past the magnets, the effects will be still further increased.

The knowledge of these facts led to the construction of the various magneto-electric machines, of which a familiar type is seen in those small ones used for medical purposes. They contain a large horse-shoe magnet, close to the end of which two bobbins of copper wire are made to revolve at a high speed, and all who have used these machines know that the more quickly they turn the handle the greater shock the person receives who is being operated upon. The current generated is really very feeble, the shock being produced by interrupting it at every half revolution by means of a small spring or other suitable mechanism. If the current is not so interrupted, it cannot be felt at all, which may be proved by lifting up the spring on the spindle of the ordinary kind. The current is an alternating one, and changes its direction throughout the circuit, however extended it may be, at every half revolution. If it is required to have a continuous current, use must be made of what is termed a commutator, and I shall endeavour to explain the manner in which it acts as simply as possible. Without going into any further details as to the construction of the bobbins, and their action at any particular moment, I shall content myself with saying that if the wire on the two bobbins is continuous, and the ends are connected, the current will flow one way during half a revolution, and the other way during the other half. Now, in Fig. 2, on the

Shock produced by interrupting of current.

The current must be commutated.

Description

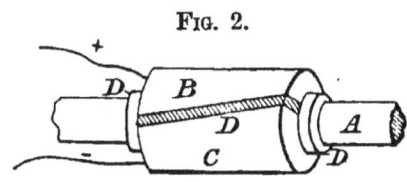

FIG. 2.

spindle A on which the bobbins are fixed, is fitted a split collar formed of two halves B and C, to which are joined respectively the ends of the wires + and −.

of commutator.

This collar is insulated from the spindle by a suitable insulating material, that is to say, a material which does not conduct electricity, such as wood, ivory, &c., and is represented in Fig. 2 by the dark parts D. So far the circuit is not complete, so that however quickly you turn the machine no current is produced. If, however, some means is employed for joining B and C by a conductor, the alternating current is produced as before. In Fig. 3, I show a

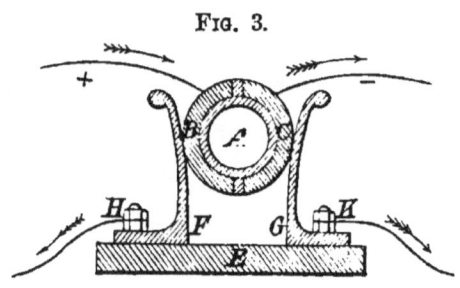

Fig. 3.

section through B A C. On a base E made of wood, are fixed two metal springs F and G, which are made to press against B and C respectively; wires are connected at H and K, which, joined together, complete the circuit. A continuous current is said to be + or positive where it leaves a battery, and − or negative where it returns; it will be convenient to use these signs and terms in the following explanation. At one portion of the revolution the spindle will be in the position shown in Fig. 3, and the + current is flowing into B, through F, to the terminal H, thence through the circuit to the terminal K, through G to C, and so back through the − wire to the bobbins of the machine. In Fig. 4 the spindle has made a half revolution, bringing B in contact with G, and C with F. But by this half turn the current is reversed in the bobbins, and the + current flows into C, through F, to terminal H as before, and through the circuit to K, through G and B, back to the bobbins. Thus you see that in the circuit the current will be always in the same direction, or

Current though alternating in the dynamo, is continuous in the circuit.

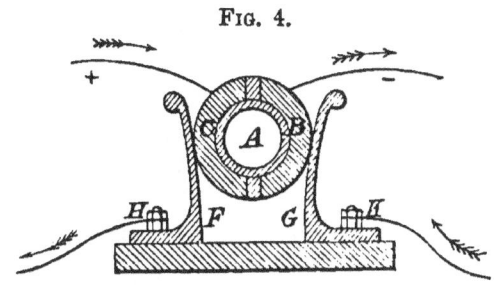

Fig. 4.

continuous, although in the bobbins it is alternating, and may be used for any purpose for which a continuous current is required, such as electroplating, &c.

There are various forms of the magneto-electric machines, as well as of commutators, but the foregoing shows the general principle of them all.

Continuous current used for electroplating.

Dynamo-electric Machines.

It will now be necessary to explain the nature of a dynamo-electric machine, called, for shortness, a

10   WRINKLES IN ELECTRIC LIGHTING.

dynamo, and to show in what it differs from a magneto-electric machine.

I have explained how an electric current is produced by a wire passing in front of a magnet; now, this magnet may either be of the ordinary kind, or it may be what is termed an electro-magnet. One of the effects which electricity can be made to produce is the magnetising of steel bars to form the ordinary and well-known permanent magnets which are used in ships' compasses, &c. To produce this effect, part of the wire in a circuit is made into a spiral as in Fig. 5.

<small>Current will magnetise an iron or steel bar.</small>

Fig. 5.

The steel rod to be magnetised is placed within the spiral, and a continuous current of electricity is then sent through the wire, which causes the rod to become magnetised with a North pole at one end, and a South pole at the other. The more current is passed through the circuit, and the more turns are in the spiral, the more quickly and strongly is the rod magnetised; and it will retain its magnetism for an indefinite time if made of suitable steel. There is a point at which the metal is said to be saturated with magnetism, and the strength it has then acquired will be that which it will retain afterwards, although while under the influence of the current that strength may be considerably exceeded. If instead of a steel

<small>Permanent magnet.</small>

<small>Electro-magnet.</small>

rod one of iron is placed in the spiral, and the current is passed through as before, it will be magnetised in the same manner; but as soon as the current is stopped, the rod loses almost all its magnetism, and if the current is then passed in the opposite direction the rod will be magnetised in the opposite way. The softer and more homogeneous is the iron, the more instantaneously will it acquire and lose its magnetism, and the greater strength of magnetism it is able to acquire. An iron bar, round which are wound a large number of turns of insulated or covered wire, constitutes an electro-magnet. The difference then between a magneto-electric and a dynamo-electric machine is, that in the former permanent magnets are used, and in the latter electro-magnets take their place. I do not intend to go into particulars as to the construction of the various dynamos in present use, as there are many books to be had in which these machines are fully described. I need merely say that in the so-called continuous-current dynamos, the whole or part of the current produced is made to pass through the coils of the electro-magnets, thus inducing in them the required magnetism. I showed how, in the magneto-electric machine, the currents are collected by means of a commutator, and it is evident that in Figs. 2, 3, and 4 there might be separate wires coming from each bobbin to B and C; and if there were more than two bobbins, there might still be

Where the magneto and dynamo machines differ.

two wires from each to B and C. On the other hand the collecting collar might be split into more sections; in fact there might be as many sections as bobbins. To show how the current is collected in continuous-current dynamos, I must give a short explanation of the revolving part or armature of a standard type of machine.

Armature of so-called continuous-current dynamo.

In Fig. 6 is shown a horse-shoe magnet, with its North and South poles, N and S. Between these poles is made to revolve the armature, composed of a number of coils of wire made to form a ring like a life-buoy. The ends of the wires are made to lie along a collar on the spindle, made of some insulating material, each wire being parallel to its neighbour, and kept separate from it, as shown in Fig. 7.

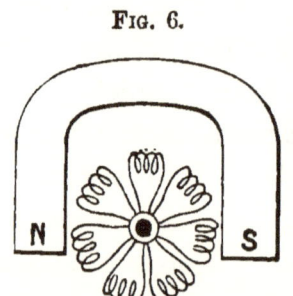

FIG. 6.

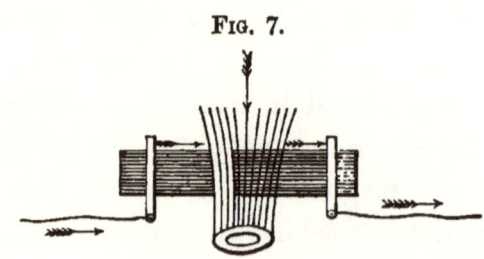

FIG. 7.

Type of commutator.

These wires are so arranged that if one end of a sectional coil is on top of the spindle at a given moment, the other will be on the under side. If then, as shown in Fig. 7, a rubber of copper, made in the form of a brush of copper wire for convenience, is placed in contact with the upper part of the commutator collar, and another similar one with the lower, it is evident the circuit will be completed in the same manner as before explained. *Commutator brushes.*

Fig. 8.

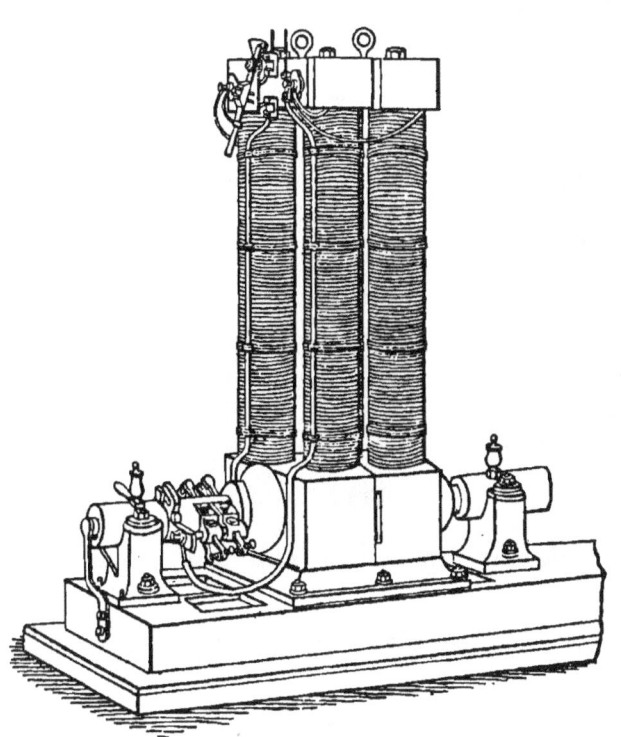

Edison Dynamo.

A wire which is + when above the spindle, will be − when below it, and as the spindle revolves the current changes in the various wires from − to +

14 WRINKLES IN ELECTRIC LIGHTING.

Current continuous in the circuit.

as they reach the top, so that it will always therefore be + in the upper brush and − in the lower one, and will accordingly be continuous through the circuit. It will be seen in the illustrations of various continuous-current dynamos, that though their shape and arrangement differ, the mode of collecting the current is much about the same as I have described above. Figs. 8 and 9 show some of the continuous-current dynamos at present in use.

Fig. 9.

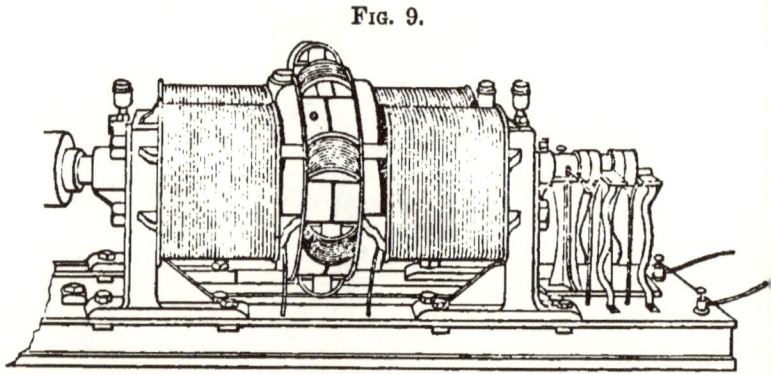

Brush Dynamo.

Alternating-current dynamos.

I will now explain the nature of an alternating-current dynamo.

The principal difference between the continuous- and alternating-current dynamo, is in the number of magnets used. Most of the former have only four magnets, while the latter have frequently as many as thirty-two. In reality, as I have shown, these are all alternating-current dynamos, only that in the so-called continuous-current ones, the current is commutated, whereas in the others it is not, but is used as it is produced. In the principal alternating-current dynamos, a number of small magnets, usually sixteen, are attached to a framework directly opposite a similar number of others of the same size, the space between the ends being only about an inch or two. These are all electro-magnets, and are wound in such manner that when excited by a current, every alternate one shall have the same magnetism, as in Fig. 10, and every opposite one a contrary magnetism. This produces an intense magnetic field between the ends of the magnets, and in this space revolves the armature. This armature, in the Siemens dynamo, is composed of a disc having as many bobbins on the periphery as there are magnets on each side of the dynamo. As each bobbin approaches each magnet a current is induced in one direction, which is reversed when the bobbin recedes; thus an alternating current is produced, which is collected by connecting the ends to insulated rings or collars on the spindle, and having small copper brushes or rubbers in contact with them. In the Ferranti dynamo, the armature is quite different, and much more simple, as comparison of Figs. 11 and 12 will show.

It consists of a copper tape bent in and out so as to form a sort of star with eight arms, the number of

Fig. 10.

layers of insulated copper tape being from ten to thirty, according to requirements. The centre is made in a similar shape with bolts or rivets holding each convolution in place. The two ends of the

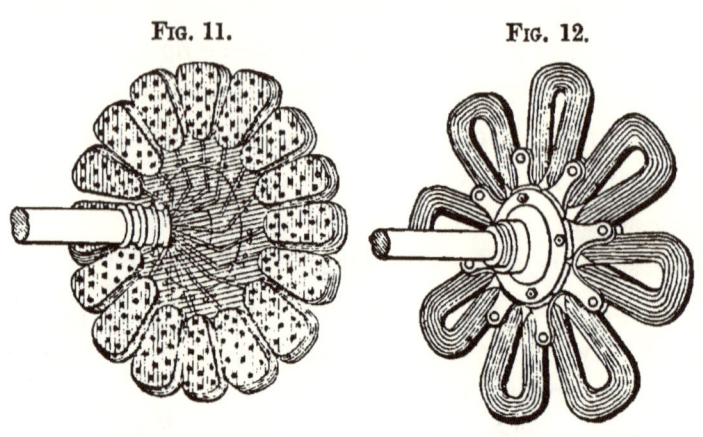

Fig. 11.   Fig. 12.

Siemens Armature.   Ferranti Armature.

tape are attached respectively to two collector-rings on the spindle, against which press two solid metal rubbers which carry off the current for use in the circuit. It can be shown that as each arm approaches a magnet a current will be induced in one direction, which will be reversed as each arm recedes; and therefore an alternating current will be produced. As there are sixteen magnets for the armature to pass at each revolution, there must be sixteen alternations of the current during the same time, so that if the speed of the armature is 500 revolutions per minute, there will be 500 × 16 = 8000 alternations in one minute. These alternations being so ex

Large number of alternations of the current.

tremely rapid, when this current is used for electric lighting, the steadiness of the light will be in no way affected, but will remain as constant as with a continuous current.

Fig. 13.

Siemens Alternating Dynamo.

The alternating current produced by these dynamos cannot be used for exciting an electro-magnet, as the magnetism would be reversed at every alternation; a separate small dynamo of the continuous type is therefore used as an exciter to magnetise all the electro-magnets in the field, and it is usually coupled on to the same spindle, and therefore goes at the same speed as the alternating-current dynamo. The exciter is usually of a size to be able to do alone

Alternating current cannot be used to excite an electro-magnet.

Exciter coupled on to same spindle as dynamo.

18   WRINKLES IN ELECTRIC LIGHTING.

Power of exciter if used alone.   about one-tenth to one-twentieth of the work that the larger machines does in the way of lighting; so that if from any cause the latter is disabled while the ship lighted by it is at sea, the exciter may be used alone to do a portion of the lighting, in the first-class saloon for instance. This can only be done if the exciter is so constructed as to give the proper E.M.F. that the lamps require.

FIG. 14.

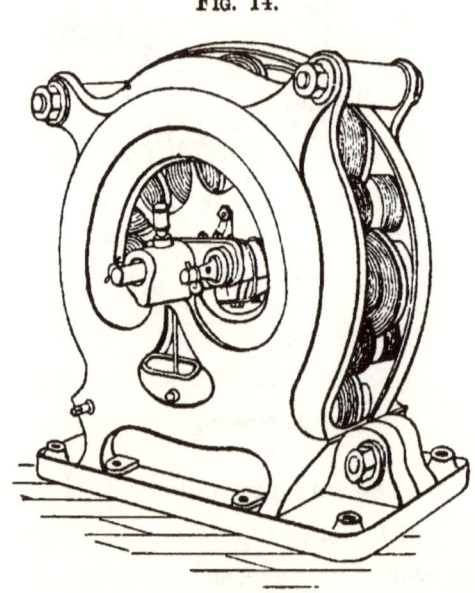

Ferranti Alternating Dynamo.

Figs. 13 and 14 are illustrations of two of the alternating current dynamos in use on board ship and elsewhere.

Electric Lamps.

I have explained how power can be converted into electric currents, either continuous or alternating, and I must now show how these currents can be applied to the production of light. Production of electric light.

The current may be used to produce an *arc light* in the following manner:—Two carbon rods, A and B, are held by suitable means in the position shown in Fig. 15, and the two wires from a dynamo are joined respectively to A and B, the upper one always being the positive lead when a continuous current is used. When the current is sent through the circuit, it passes through the carbons A and B, which are conductors. Immediately this occurs, suitable mechanism in the lamp, being acted on by the current, or by hand in the case of search-lights, or by clock-work, moves the two carbons a small distance apart, with the consequence that a dazzling arc of light is formed between them. If the carbons get too far apart, the mechanism brings them nearer together again, and on the delicacy with which it acts, depends the steadiness of the light. It would be useless to explain how this mechanism acts, as it is in a different form in each maker's lamp. Some lamps have been constructed for use with an alternating current, but with the majority a continuous current is used. Arc lights.

Fig. 15.

Mechanism to regulate carbons.

Some lamps suitable for alternating current.

20 WRINKLES IN ELECTRIC LIGHTING.

While an arc light is burning the carbons waste away, the upper one more rapidly than the lower, and the mechanism has to approach them constantly to make up for this waste.

When carbons are consumed light goes out.

When the carbons are consumed as far as convenient, an automatic arrangement cuts off the current, and the light goes out; or it diverts the current to another set of carbons, which at once light up. The carbons are made in suitable lengths to last a certain number of hours, four, six, eight, &c. In Fig. 16 is shown an arc lamp complete.

Arc lamp very complicated.

An arc lamp is of necessity a complicated affair, which it is not advisable to have on board ship, except where an electrician is engaged permanently.

Jablochkoff candle.

Another way of producing light is to use the current in what is called an *electric candle*, of which a familiar type is the Jablochkoff candle.

Fig. 16.

Arc Lamp Complete.

Fig. 17 shows the form of this candle, A and B being two carbon rods parallel to one another, and joined, but at the same time insulated from one another by kaolin, a sort of chalky substance, which is a non-conductor.

The wires C and D from the dynamo are joined respectively to A and B through metallic supports, as in an arc lamp, and when the current is turned on it flows through C A and across by a small strip of carbon E to B and D back to the dynamo. The strip E is only large enough to carry the current across for a moment, and is immediately consumed, but an arc of light is then formed between the carbons as in the arc lamp. As the carbons consume, the kaolin in between burns away, just in the same manner as, in an ordinary candle, the wick is consumed and the wax melts and burns away, except that in the latter case the wax feeds the light, whereas the kaolin is only used to keep the carbons the required distance apart and the arc of light from running down them. It is evident that the carbons must be consumed equally, for which reason use must be made of the alternating current. Any unsteadiness that occurs in the light produced is consequent on unsteadiness of the current, or impurities in the carbons, &c., there being no mechanism of any kind required. These candles do not give such a great light as arc lights, but it is of the same nature in every way. Fig. 18 shows one of these candles in its holder, from which can be

FIG. 17.

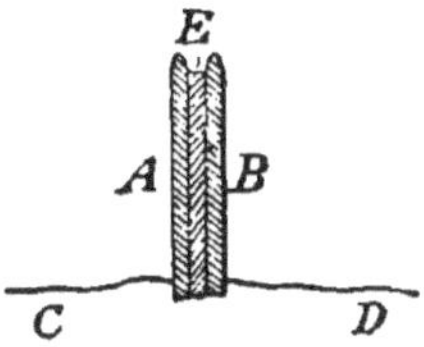

Arc formed between the carbons.

Candles require alternating current

seen how electrical contact is made with the two carbons.

If the current is interrupted in any way, and the light goes out, it will not be produced again auto-

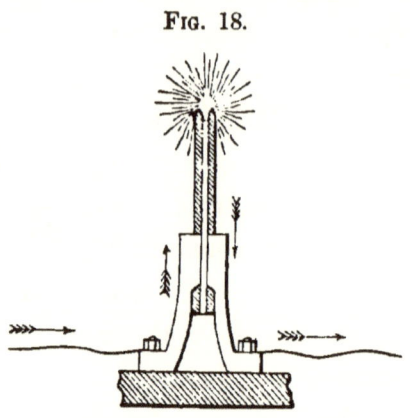

Fig. 18.

matically, but requires a small piece of carbon between the two carbons as a path for the current to pass across as in the beginning.

Incandescent lamps.

A third form of electric light is produced by using the current in an *incandescent lamp.*

To explain the action of an incandescent lamp, I must refer back to what I said about wires getting heated by a current being passed through them which was too large for their capacity. If two large wires are joined by a small one, and a strong current is passed through the circuit, the small wire rapidly gets red hot, and finally fuses. If this small wire is

Vacuum formed in lamp

contained in a globe from which the air is exhausted, when the current is passed through it, it gets red, then white hot, and when very brilliant gets fused. *prevents combustion.* If, instead of wire, we have in the small globe a thin filament of carbon, when the current is passed through, we get a brilliant light which remains constant because the carbon does not fuse, and it cannot burn away for want of air. Fig. 19 shows a Swan lamp, and Fig. 20 an Edison lamp, both made on this principle.

FIG. 19.

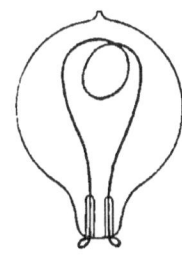

FIG. 20.

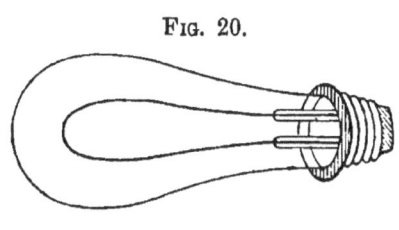

If in these lamps the vacuum were perfect, the carbon filament would never get consumed; it is, however, impossible to get a perfect vacuum, but the better it is, the longer will the filament last. Incandescent lamps are the only ones that are suitable for house or ship lighting. They give a yellowish light like a good gas-flame, they do not consume the air of a room, they cause no smell, and only give out a very slight heat. They are perfectly safe, because if the globe gets broken and allows air to get in, the filament is instantly consumed, and the light goes out. They can be put in all sorts of places where it would be impossible to

Vacuum not perfect.

Advantages of incandescent lamps for house and ship lighting.

have any other lamps, such as near the ceiling, close to curtains, in a room full of explosives or combustibles, and even under water. They are not affected by wind; they can therefore be used under punkahs, or near open windows, sky-lights, or ports, or in the open air. These lamps can be used with either continuous or alternating currents, but will probably last longer with the latter, because, when a continuous current is used, particles of the carbon of the filament appear to be conveyed from one end of the filament to the other, reducing the thickness at the one end, until finally it breaks. This evidently cannot occur with an alternating current, as the impulse in one direction is counteracted by the following one in the opposite direction. If the current used is of too high a tension for the lamps, they will show an intensely brilliant light for a short time, but the filament will soon be destroyed, and the lamp rendered useless.

Unaffected by wind, and suitable for either continuous or alternating currents.

LEADS.

We have now to consider the means used for conveying the current, continuous or alternating, to the lamps we intend to use. The leads for the electric current, which correspond in some measure with the pipes which convey gas, are made of copper wire, as pure as can be obtained, covered with some insulating material to prevent the escape of the current through contact with other conductors. The

Leads made usually of copper wire.

size of the wire is regulated according to the amount of current which is to be conveyed; it will do no harm to have it of twice the required section, but if it is of less than the required section, it will offer so much resistance to the passage of the current, that it will probably get fused in a very short time. If the lead attached to one terminal of the dynamo comes back to the other terminal without there being any lamps in the circuit, or other means of making use of the current, it is said to be short circuited, and if the dynamo is kept going something must give out very soon. The two leads must therefore never be connected with one another, except by a lamp or other resistance, and the manner in which the lamps are placed, and the size of the leads, depend upon the relative tension and quantity of current and the kind of lamps to be used. If the current is to be used in arc lamps it is usual to have a high E.M.F., which allows of the leads being of small section; but if it is to be used in incandescent lamps it is found more convenient to have a low E.M.F., and as this implies a large quantity of current, the leads have to be of large section. *Short circuit. High E.M.F. for arc lights, but low for incandescent.*

Arc lamps usually require to be placed in series, that is to say, in such a manner that the current, after leaving the dynamo, passes through each lamp in succession. The E.M.F. required in this case is the sum of the E.M.F. for each lamp, the quantity required being the same as for one lamp. This *Arc lights in series.*

accounts for the high E.M.F. used in arc lighting and the small size of the wire for conducting the current. Incandescent lamps can be either in series or parallel, and frequently the two systems are combined. To explain the meaning of having lamps parallel, we will suppose the two leads from a dynamo to be taken along a wall, parallel to one another, and about six inches apart, ending at the end of the wall, but not connected in any way. If we then place lamps at intervals between the two leads, connecting one loop of each to the upper lead, and the other to the lower lead, by means of small copper wire, these lamps are said to be all parallel. In this arrangement the current required is the sum of the quantity necessary for each lamp, but the E.M.F. is the same as that required for one lamp of the same kind. As we therefore require to send a large quantity of current through the leads at a small pressure or E.M.F., these leads must be of large section. In the above arrangement each lamp may be turned on or off separately without affecting the others. Sometimes two or more lamps are placed in groups between the parallel leads; these are then in series with regard to one another, and can only be turned on or off two or more at a time, in other words, one group at a time. If our dynamo is producing a current of 100 volts E.M.F. when working at its proper speed, and our lamps are 100-volt lamps, we shall be able to turn each lamp on or off separately; but if we want to put in 50-volt lamps, we must place two together, and we shall then have to turn them on or off two at a time. I am supposing that in both cases the lamps require the same quantity of current, though of different E.M.F.

To prevent the lamps being spoilt by the current being too strong through a sudden increase in the speed of the dynamo, as also to prevent the leads getting fused, and perhaps setting fire to the casing, it is usual to have safety fuses in various parts of the circuit. These are of different kinds, but a typical one consists of a small lead wire, large enough to carry the normal current, but which fuses when the current is too strong, and at once interrupts its passage. The lamps in the same portion of the circuit are then extinguished and so saved from destruction, and cannot then be lighted again until the fuse is renewed, which, however, can be done with ease. *[Safety fuses.]*

Ship Lighting.

We will consider now the case of a steamship to be lighted by means of incandescent lamps. It is sometimes a matter of some difficulty to fix on a suitable position for the dynamo and engine, especially in ships which have already been running for some time.

In selecting a position, it must be borne in mind *[Position for dynamo.]*

that a dynamo will work best in a cool clean place, cleanliness being most important. If a lot of coal dust is flying about where the dynamo is working, it will be drawn into it, and tend to impair its electrical, as well as mechanical efficiency. If the dynamo is kept properly lubricated, it will work well enough in a hot place, but we must remember that the heating of the wire which makes up a large portion of the dynamo, reduces its conductivity, so that the cooler it is kept the better. The dynamo should be so placed that the engineer can get to every side of it easily. If a quick-speed engine is to be used for driving it direct, it will make a very compact installation, but there seems to be some difficulty as yet in getting suitable reliable engines, besides which many marine engineers object to quick-speed engines altogether. If a slow-speed engine is to be used, a belt is of course required to get the necessary speed on the dynamo, and various precautions are needful to prevent the belt slipping off the pulley when the ship is rolling about in a sea-way. In all cases, the engine and dynamo should be placed with their spindles fore-and-aft, or in a line with the ship's keel, the rolling being felt more than the pitching. There are various ways of keeping the belt from slipping off the pulley. Some have flanges on the pulley, others have guides or rollers on each side of the belt, each plan having its advantages and disadvantages; but some plan must be used, otherwise the belt slips off, usually in the middle of the first-saloon dinner, and out go all the lights, besides which the belt may be considerably damaged before the engine can be stopped. The engine must be one that will work very steadily, otherwise the lights will pulsate at each revolution of the engine, which is most unpleasant. If the engine is a single one, it must have a large flywheel, or a driving-wheel large and heavy enough to answer the same purpose. The engine requires a good sensitive governor, so as to keep the speed regular when some of the lamps are turned on or off. When the engine and dynamo are in the main engine-room, the throttle-valve, or a stop-valve, should be in a convenient place for the engineer on watch to get at so as to instantly shut off the steam if the belt slips off or breaks. In ships where an electrician is carried there will not be the same necessity for this precaution. It is necessary to have some means of tightening up the belt, so as to keep it from slipping round the pulley. Where the engine and dynamo are on the same level there may be a screw arrangement in the base-plate of the latter by which the distance between centres can be increased. Where the engine and dynamo are on different levels, and the latter is a fixture, recourse must be had to a roller, bearing against the upper part of the belt and capable of screw adjustment. If link leather belting is used, it will be found

Engine must work steadily.

A good sensitive governor wanted.

The belt must be kept tight.

necessary to take out several rows of links each day until it has finished stretching. A very handy thing to use for this purpose, and which can be made on board by an engineer, is a double clamp with a screw in between, just like the ones which are being sold for stretching trousers which have got baggy at the knees. Whatever belt is used, it is very important that there should be no joint or inequality which can cause a jump or slip when going over the pulley, as this will cause the lights to pulsate each time. In America friction gearing has been tried, but I do not know with what success. From my experience of friction gearing, I am inclined to think it might do very well. There is certainly no doubt that direct-acting quick-speed engines are the ones to use, and it is only a question of getting a suitable one.

The dynamo being firmly fixed in position, the main leads are connected to it, and carried along to the switch-board, which should be in a convenient position near at hand. On this switch-board are usually placed the large safety fuses. The board should have a cover to it, to prevent any one meddling with it, and to keep it clean. The main leads are of a large size, and from these other smaller ones branch off, being spliced and soldered to them. It is a very good practice to use leads of two different colours, as we can then work by the following rule: Never connect together two leads of different colours except by means of a lamp or other

A handy belt stretcher.

Friction gearing.

Switch-board near dynamo.

Leads of different colours.

WRINKLES IN ELECTRIC LIGHTING. 31

resistance. The size of the various leads depends on the current to be conveyed, and is a matter for the electricians. On the main-deck of a large passenger steamer, the main leads may be carried along side by side under the upper deck, and from these, smaller ones branch off into the various sets of rooms, smaller ones still going into each room. In each room there will be one lamp with its switch to turn it on or off as desired, and a safety fuse. The lamps are held in small brackets, and are contained when desired in frosted globes, which diffuse the light and make it very pleasant. When these globes are held rigidly in the brackets, the least knock breaks them. A very good bracket I have seen in use is one which allows the globe to move about on its support when touched, being at the same time sufficiently a fixture to resist the motion of the ship; and in the particular ship in which I saw these used in the first saloon, there was not a single breakage during a four months' voyage. The switches for turning each light on or off can be under the control of the passengers, or, on the other hand, they can be fitted to work with keys kept by the stewards, as thought most desirable.

Main leads and branch leads.

Lamps held in frosted globes.

Switches for each lamp.

The lamps used can be of various candle-powers, within certain limits, and of whatever make is considered best. They can also be of various makes, as long as they are constructed to stand the same E.M.F. The lamps in the passenger berths give

Lamps of various candle-powers.

quite sufficient light if of 10-candle power; the ones for lighting the saloons, passages, and other large spaces, may with advantage be of 20-candle power. In these days of luxurious travelling, when the various lines are trying to attract passengers to their particular ships, what follows may be thought worth consideration. In steamers going through the tropics to India, China, Australia, &c., it is usual to get up dances, concerts, and other entertainments on the quarter-deck, at times when it would be impossible to do anything below on account of the heat. The quarter-deck then has to be lighted up. This is effected by means of globe oil-lamps hung about here and there, two being hung in front of the piano, in unpleasant proximity to the head of the obliging lady pianist. Now in a ship lighted by electricity, there is no reason why a couple of leads should not be brought up from below through a skylight or other opening, on to the quarter-deck. Indeed the leads might be arranged to screw into a place in the deck, or on the side of a skylight, just in the same manner as a hose is connected for washing decks. These leads would have holders for lamps fitted permanently at intervals, and when required for use would be stopped up along the awning-spar or ridge-chains, and the lamps screwed or hooked into the holders. With a few handy men, five or ten minutes would suffice to arrange the whole thing after the leads had once been fitted. The leads once fitted for this purpose would be always ready for use, and could be kept coiled away in a box which might also have a compartment to contain the dozen or so of lamps required.

[margin: always ready, and easily fixed up.]

If the dynamo is already running as many lamps as it is capable of, some of the bedroom lights may be turned off while the quarter-deck is being lighted. Another thing which I think has not yet been done is the following. When working cargo at night, and indeed during the day to some extent, lights are of necessity used in the holds. The *theory* is, that no naked lights shall be allowed, but the *practice* is this: lamps are taken below, get knocked about, the wicks fall down and want pricking up, the lamps are opened for this purpose, and as they are found to give more light without a dusty glass round them than with it, they are left open. Candles are often taken below lighted, and even matches struck to see the mark on a bale. I am aware that arc lamps are used in the Royal Albert Docks, London, in connection with the dock lighting, lamps being carried below when required, with flexible leads attached, and that, in some few steamers, arc lamps have been used in the same manner in connection with their own plant. These arc lamps are, I think, not nearly as suitable as incandescent lamps for the purpose of lighting up a ship's hold; the light is too glaring, and casts deep shadows amongst the bales and cases, besides which, the lamps are large and clumsy. I

[margin: Lighting of ships' holds.]

[margin: Danger of fire with oil lamps.]

[margin: Arc lamps not suitable.]

Arrangement of leads for incandescent lamps.

would suggest that leads should be carried behind the stringer-battens in the ship's side, or along under the next upper-deck, having simple sockets or holders for incandescent lamps at certain intervals. Whoever might be in charge of the hold would screw or hook on the lamps as required, and so light up every part of the hold thoroughly while work was going on. There would be no risk of fire, and I am

Work carried on better, and pilfering of cargo prevented.

convinced that the extra leads and lamps would pay for themselves in a very short time, because work would get on more quickly, and pilfering of the cargo would be in a great measure put a stop to. The leads for the holds could be so arranged as to be

Hold leads disconnected while at sea.

quite unconnected with the dynamo while at sea, so that there could not be the remotest possibility of the current finding its way below when not wanted. In fine, there is no reason whatever why a ship's hold should not be lighted up when required, as well as a warehouse or store on shore.

Installation complete.

Now, we will suppose that our installation is complete, ready for working, everything having been pronounced in order by the electrician who has

Lights wanted as night approaches.

looked after the work. Evening is approaching, and the lights will soon be required; we must therefore see that our engine and dynamo are ready for a

Precautions before starting dynamo.

start. If the engine and dynamo are separate, the belt must be felt, to see that it is tight enough, otherwise it must be tightened by whatever means are provided for the purpose. We must also see that the engine and dynamo are properly oiled, and that the worsteds are down the tubes of the oil-cups, and working properly, not dry, as I have known them to be, with fatal results to the dynamo. If the lubrication is performed by means of tubes leading to each bearing from an elevated oil-box, we must see that the oil really gets to the bearings, and regulate its flow as required. The commutators and collector-rings and rubbers require only a wipe of oil, just sufficient to prevent undue wearing of the surfaces; if too much is put on them, they will spark a great deal, and sparking will wear them away more quickly than friction. The brushes of copper wire which collect the current of the exciter dynamo, and others of similar pattern, must be placed so that the ends press on the commutator as shown in Fig. 21. The ends should project just a little way beyond the point or line of contact, and when the dynamo is running, there should be very little sparking. I am supposing that our plant con-

Lubrication must be perfect.

Commutators and collectors require very little oil.

Position of brushes.

Fig. 21.

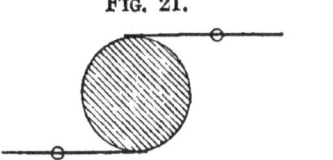

Fig. 22.

sists of an alternating-current dynamo with a small exciter. The wires leading from the exciter to the other dynamo remain always connected, as there is no need for meddling with them.

We will now start the engine, and thereby set the dynamo going round, slowly at first, and gradually up to the speed required. The main switches are not yet turned on, so there is no current going through the leads as yet; what then is being done? A current is being produced by the exciter only, and is magnetising the electro-magnets of the larger dynamo, and if we want to know if it is really doing its work as intended, we just hold a small pocket-compass over the ends of two opposite magnets of the dynamo, and observe how the needle points. It should at once take up the position shown in Fig. 22, and if then held over the next couple in like manner, the needle should simply turn round, and point in exactly the opposite direction. If it points in any other direction, there is something wrong with the connections. If, however, the connections are right at starting, they will of course remain right, and there should be no need for this test. It is well to remember that when dynamos are working, they are, or contain for the time being, very powerful magnets, therefore if we bend over them to examine them, our watches will get magnetised, which does not improve their qualities as time-keepers. Say that our dynamo is now going round at the required speed, which may be 500 or 600 revolutions per minute; the engine is not using much steam as yet, because very little work is being done. We now switch on a set of

WRINKLES IN ELECTRIC LIGHTING. 37

lamps; this closes the circuit, and the large dynamo begins to produce its alternating current, which goes through the lamps and lights them up. This, however, gives the engine more work to do, and more steam must be turned on, otherwise the necessary speed will not be kept up. We switch on all the other lamps as required, and must see that the speed of the dynamo is kept constant. A difference of a few lamps, affecting the engine to a small extent only, should be compensated automatically by the governor. If the brightest lamps are not bright enough, the speed should be increased a little, but care must be taken not to overdo it, because if the current is too strong, some of the safety fuses will melt, and the corresponding lamps will go out. It must not be inferred from what I have said, that it is necessary to run the dynamo at first without switching on any lamps. On the contrary, a better effect will be produced if all the lamps are switched on before starting, as they will then gradually work up to their full brilliancy; whereas, if one set of lamps is started first, and run bright, and we then switch on another set, the current at first will be too small for the two sets, and the first set will get quite dull, remaining so until the dynamo is going at its proper speed again. When lighted up for the first time, it will be found that some of the lamps are much brighter than others; this is because the lamps at present made

Marginal notes: Current is produced in large dynamo. Difference of a few lamps compensated by governor. Turn all lamps on, and light up gradually. Inequality of light in different lamps.

are not of perfectly equal resistances. We must go round, then, and note where the dull ones are, and we can either at once, or during next day, shift them into the bathrooms and places where such a perfect light is not required. All the lamps in one room, the first saloon, or music room, for instance, should be equalised as much as possible, and in such places the brightest should be used. Nothing looks worse than to see a couple of dull lights in the same room as a lot of bright ones. By seeing to these matters we can make the lighting much more satisfactory than it otherwise would be. During the first few evenings we shall probably have some of the lamps go out through the filaments breaking.

Weeding out of bad lamps. This I consider a weeding out of defective lamps, because if it were that the current was too strong, the fuses would have given way. Some of the fuses give way when the current is *not* too strong; this is owing to imperfections in the fuses, and they must be replaced by spare ones. For the sake of economy,

Lamps not to be run too bright. it is well not to run the lamps too bright. Without giving the lamps the maximum current a very good light can be obtained, and they will last all the longer. I need hardly say that there is a medium in this as in everything else, and it does not look well when a candle is placed alongside of an electric lamp to enable a person to read or write in comfort.

All this time the dynamo is running, and we must feel the bearings occasionally to see if they are keeping cool. There will be no trouble if the lubrication is all right. If the oil does not get into the bearings as it should do, they will heat, jam the spindle, or seize, and bring up the engine or break the belt. The lights will then all go out, and everybody will say hard things of the electric light, while the fault really rests with us. Sometimes seizing occurs through the spindle not being slack enough in the bearings, but this generally occurs while testing the dynamo at the works. *(No trouble with dynamo if oiling is attended to. Seizing.)*

It must be borne in mind that in dynamos the spindle must be a good fit, and there may be room in the bearings for ordinary engine-oil while there may not be for a thicker oil, such as castor oil. Therefore, if the bearings show a tendency to heat, it may improve matters to thin the oil used with petroleum. While giving the dynamo its proper supply of oil, we must only apply it in the proper places. If we let the bobbins get smothered in oil, the insulating material on the wire will get rotted, and a short circuiting will probably take place. The dynamo cannot be kept too clean, and there should be a canvas cover to put over it while not in use, especially while coaling. We will suppose that all is going on right; a steward comes along and says: "Mr. So-and-so, I cannot get the lamp in number 6 berth to light although I have turned the switch the right way." "All right, I will go and look at it," you answer. Now, let us see what is the matter. We unhook or *(Oil must be thin. The dynamo must be kept clean. Little troubles with the lamps.)*

unscrew the lamp, and look at the filament; it is not broken. We replace the lamp again, and are careful that it makes good contact; but still no light. Let us look at the safety fuse; why, there is none! it has been missed out. We get one of the spare ones out of our electric store, and put it in its place, and the lamp lights properly at once. We find another lamp out, and look at it. We see at once that the filament is broken, so there is no question about this one; it must be changed. Hallo! what is up with this one? it goes in and out all the time like a flash light. The current must be getting to it all right, otherwise it would not light at all. I see what it is; it is a Swan lamp, and the spring is not pressing quite fairly on it, so that one hook is making good contact, while the other tightens and slacks with the vibration of the ship. This is soon set right by turning the spring round a little, or hooking the lamp the other way. Or it is an Edison lamp, which has got slightly unscrewed, and no longer makes good contact at the back end of the holder. In some lamp-fittings the ends of the leads are held in a spring grip in the base of the bracket, and it may happen that they have slipped out, and so broken the circuit, and extinguished the light. In the Swan lamps, and others of a similar pattern, one of the little platinum loops in the base of the lamps sometimes gets broken off; the lamp is then of no further use. To recapitulate, if a lamp goes out, the first thing is to see if the filament is broken, next if it makes good contact. If it does not then light up, see if there is any current getting to it; this may be found out by touching the two hooks in a Swan holder, or the back and side of an Edison screw holder, with a moistened finger. With a current of 50 volts a slight tickling sensation will be felt if the current is passing through. If this cannot be felt, there must be some part or other disconnected, perhaps the safety fuse has given out, or the ends of the leads got adrift from the bracket. If in any doubt about the lamp, try another in the same place.

Recapitulation.

A current of 50 volts is hardly felt.

In some steamers incandescent lamps are used in the side lamps; they can easily be fitted for this purpose, especially when the ship is provided with lighthouses built in, as in the Anchor Line steamers. Two or more incandescent lamps can be arranged on a small stand, which will slide into the lantern, taking the place of the regulation oil lamp, and connected by flexible leads to the other leads. It would be easy to put six 20-candle power lamps in a group in each lantern, as it does not matter in what position they are placed; two might be used on ordinary occasions, while on a foggy night, the whole six could be switched on. If one lamp went out through the filament giving way, it would not affect the others, so that there would still be a light in the lantern. If, through some breakdown of the engine or dynamo, the electric current were no longer to be had, then it

Incandescent lights for side lights.

would only be necessary to withdraw the stand of lamps, and put in the ordinary regulation oil-lamp. The mast-head lamp could also be fitted with the electric light, as indeed has already been done. On no account, however, should an arc light be used, as besides being too dazzling, it is much too uncertain; in fact many other reasons could be given for rejecting it. It is even a question whether it is an advantage to have incandescent lamps for a mast-head light. There is certainly the great advantage of not having to pull the lamp up and down to trim it, a rather risky performance in heavy weather, and also of the light not being affected by any wind that may get into the lamp; though as regards the first, English officers would never be satisfied to see a lamp dangling on the stay all day long, as appears to be the custom in some foreign steamers, besides which it would have to be lowered to be cleaned outside.

<small>Mast-head light.</small>

<small>Arc light should never be used.</small>

The present mast-head lights are quite powerful enough already, too much so when compared with the side lights. I am not aware of any collisions having occurred through a mast-head light not being seen in time, but how many from the side lights not being seen! It was no doubt contemplated, as indeed the regulations show, that no lights should be visible about a vessel, except the regulation lights; but many who have seen a large passenger steamer go past will have noticed how her side was—one,

<small>Present mast-head lights quite powerful enough.</small>

<small>On passenger steamers,</small>

WRINKLES IN ELECTRIC LIGHTING. 43

two, or three rows of dazzling bright lights, and will have looked almost in vain for the green or red light dimly visible in the midst of all the bright ones. If bright electric lights, therefore, are shining through the ports, we must have our side lights at least as bright, so as to give them a chance of being seen. If electric lamps are used as side lights, the dynamo must be kept running all night. If it is thought desirable to put out all unnecessary lights at 11 P.M., the leads can be so arranged that these lights can be all on one or more circuits, and the necessary ones on another. *{side one blaze of light, and side lights barely visible.}*

Although the dynamo will have to go at nearly the same speed throughout the night, it will not have the same amount of work to do, and the engine will therefore use much less steam, the consumption being in proportion to the number of lights used. An economical engineer will therefore see that bedroom lamps are not kept lighted all the evening without any necessity. On shore we should never think of keeping gas-lights blazing away for no purpose, and why should we use electricity with more lavishness, especially when it is so easy to turn a light on or off. The switches might with advantage be painted with Balmain's luminous paint, and there would then be no trouble in finding them in the dark. It is well to know that on board ship, probably in all cases of electric lighting, there is no danger to life to be apprehended from touching any *{Speed of dynamo constant, but steam power used in proportion to number of lamps in use.}* *{No danger to life from electric current on board ship.}*

of the leads where bare, or indeed any part of the dynamos, as the E.M.F. is usually not more than 50 volts. It is best, however, not to try any experiments, and it is a good general rule, not to touch a bare part of a dynamo or lead with both hands at the same time. The fear of getting hurt has the good effect of keeping passengers and others from meddling with their lamps.

<small>Binnacle lamps. Electric light not suitable.</small>

I have said nothing about the use of electric lights in binnacles, though it would be a great advantage to be able to supply them with a good steady light quite unaffected by wind. There is an obstacle to their use for this purpose, in that the electric current being used near the compass, the latter is affected by it. In theory, an alternating current should have no effect; but it would require very exhaustive experiments to be made before enough confidence could be inspired concerning its innocence, and I fancy it would usually be looked upon with great suspicion by captains and officers of ships.

<small>Dynamo, if near a compass, will affect it.</small>

The dynamo being made up of powerful magnets, must of course be always at a good distance from the compasses. In some installations on iron steamers, the return leads have been dispensed with, the iron of the ship carrying the current back, in the same way that the earth or sea does it in a telegraph circuit.

<small>Notes.</small>

It is to be observed that a dynamo with *brushes* on the commutator is not necessarily a *Brush* dynamo as a good many people seem to think, the latter being named after its inventor, Mr. Brush.

A dynamo is not a *battery* as some people call it, and there is no need for multiplying names.

A pocket speed-indicator should be supplied for testing the speed of the dynamo, to see that it is kept up to proper speed, and that the belt (if used) does not slip to an unreasonable extent.

I think I have now said enough to redeem my introductory promise, and if I have, so to speak, let more electric light on to a subject previously dark to a good many people, I shall be well satisfied with my labour, and I hope that those who peruse this book will be induced to go more deeply into the subject by means of the many good books which have been written by cleverer men than I, and which enter more thoroughly into all its details.

LONDON
PRINTED BY WILLIAM CLOWES AND SONS, LIMITED, STAMFORD STREET
AND CHARING CROSS.

1888.

BOOKS RELATING
TO
APPLIED SCIENCE,

PUBLISHED BY

E. & F. N. SPON,
LONDON: 125, STRAND.

NEW YORK: 35, MURRAY STREET.

A Pocket-Book for Chemists, Chemical Manufacturers,
Metallurgists, Dyers, Distillers, Brewers, Sugar Refiners, Photographers,
Students, etc., etc. By THOMAS BAYLEY, Assoc. R.C. Sc. Ireland, Analytical and Consulting Chemist and Assayer. Fourth edition, with additions, 437 pp., royal 32mo, roan, gilt edges, 5s.

SYNOPSIS OF CONTENTS:

Atomic Weights and Factors—Useful Data—Chemical Calculations—Rules for Indirect Analysis—Weights and Measures—Thermometers and Barometers—Chemical Physics—Boiling Points, etc.—Solubility of Substances—Methods of Obtaining Specific Gravity—Conversion of Hydrometers—Strength of Solutions by Specific Gravity—Analysis—Gas Analysis—Water Analysis—Qualitative Analysis and Reactions—Volumetric Analysis—Manipulation—Mineralogy — Assaying — Alcohol — Beer — Sugar — Miscellaneous Technological matter relating to Potash, Soda, Sulphuric Acid, Chlorine, Tar Products, Petroleum, Milk, Tallow, Photography, Prices, Wages, Appendix, etc., etc.

The Mechanician: A Treatise on the Construction and Manipulation of Tools, for the use and instruction of Young Engineers and Scientific Amateurs, comprising the Arts of Blacksmithing and Forging; the Construction and Manufacture of Hand Tools, and the various Methods of Using and Grinding them; the Construction of Machine Tools, and how to work them; Machine Fitting and Erection; description of Hand and Machine Processes; Turning and Screw Cutting; principles of Constructing and details of Making and Erecting Steam Engines, and the various details of setting out work, etc., etc. By CAMERON KNIGHT, Engineer. *Containing* 1147 *illustrations*, and 397 pages of letter-press, Fourth edition, 4to, cloth, 18s.

CATALOGUE OF SCIENTIFIC BOOKS

Just Published, in Demy 8vo, cloth, containing 975 pages and 250 Illustrations, price 7s. 6d.

SPONS' HOUSEHOLD MANUAL:
A Treasury of Domestic Receipts and Guide for Home Management.

PRINCIPAL CONTENTS.

Hints for selecting a good House, pointing out the essential requirements for a good house as to the Site, Soil, Trees, Aspect, Construction, and General Arrangement; with instructions for Reducing Echoes, Waterproofing Damp Walls, Curing Damp Cellars.

Sanitation.—What should constitute a good Sanitary Arrangement; Examples (with illustrations) of Well- and Ill-drained Houses; How to Test Drains; Ventilating Pipes, etc.

Water Supply.—Care of Cisterns; Sources of Supply; Pipes; Pumps; Purification and Filtration of Water.

Ventilation and Warming.—Methods of Ventilating without causing cold draughts, by various means; Principles of Warming; Health Questions; Combustion; Open Grates; Open Stoves; Fuel Economisers; Varieties of Grates; Close-Fire Stoves; Hot-air Furnaces; Gas Heating; Oil Stoves; Steam Heating; Chemical Heaters; Management of Flues; and Cure of Smoky Chimneys.

Lighting.—The best methods of Lighting; Candles, Oil Lamps, Gas, Incandescent Gas, Electric Light; How to test Gas Pipes; Management of Gas.

Furniture and Decoration.—Hints on the Selection of Furniture; on the most approved methods of Modern Decoration; on the best methods of arranging Bells and Calls; How to Construct an Electric Bell.

Thieves and Fire.—Precautions against Thieves and Fire; Methods of Detection; Domestic Fire Escapes; Fireproofing Clothes, etc.

The Larder.—Keeping Food fresh for a limited time; Storing Food without change, such as Fruits, Vegetables, Eggs, Honey, etc.

Curing Foods for lengthened Preservation, as Smoking, Salting, Canning, Potting, Pickling, Bottling Fruits, etc.; Jams, Jellies, Marmalade, etc.

The Dairy.—The Building and Fitting of Dairies in the most approved modern style; Butter-making; Cheesemaking and Curing.

The Cellar.—Building and Fitting; Cleaning Casks and Bottles; Corks and Corking; Aërated Drinks; Syrups for Drinks; Beers; Bitters; Cordials and Liqueurs; Wines; Miscellaneous Drinks.

The Pantry.—Bread-making; Ovens and Pyrometers; Yeast; German Yeast; Biscuits; Cakes; Fancy Breads; Buns.

The Kitchen.—On Fitting Kitchens; a description of the best Cooking Ranges, close and open; the Management and Care of Hot Plates, Baking Ovens, Dampers, Flues, and Chimneys; Cooking by Gas; Cooking by Oil; the Arts of Roasting, Grilling, Boiling, Stewing, Braising, Frying.

Receipts for Dishes—Soups, Fish, Meat, Game, Poultry, Vegetables, Salads, Puddings, Pastry, Confectionery, Ices, etc., etc.; Foreign Dishes.

The Housewife's Room.—Testing Air, Water, and Foods; Cleaning and Renovating; Destroying Vermin.

Housekeeping, Marketing.

The Dining-Room.—Dietetics; Laying and Waiting at Table; Carving; Dinners, Breakfasts, Luncheons, Teas, Suppers, etc.

The Drawing-Room.—Etiquette; Dancing; Amateur Theatricals; Tricks and Illusions; Games (indoor).

The Bedroom and Dressing-Room; Sleep; the Toilet; Dress; Buying Clothes; Outfits; Fancy Dress.

The Nursery.—The Room; Clothing; Washing; Exercise; Sleep; Feeding; Teething; Illness; Home Training.

The Sick-Room.—The Room; the Nurse; the Bed; Sick Room Accessories; Feeding Patients; Invalid Dishes and Drinks; Administering Physic; Domestic Remedies; Accidents and Emergencies; Bandaging; Burns; Carrying Injured Persons; Wounds; Drowning; Fits; Frost-bites; Poisons and Antidotes; Sunstroke; Common Complaints; Disinfection, etc.

The Bath-Room.—Bathing in General; Management of Hot-Water System.
The Laundry.—Small Domestic Washing Machines, and methods of getting up linen; Fitting up and Working a Steam Laundry.
The School-Room.—The Room and its Fittings; Teaching, etc.
The Playground.—Air and Exercise; Training; Outdoor Games and Sports.
The Workroom.—Darning, Patching, and Mending Garments.
The Library.—Care of Books.
The Garden.—Calendar of Operations for Lawn, Flower Garden, and Kitchen Garden.
The Farmyard.—Management of the Horse, Cow, Pig, Poultry, Bees, etc., etc.
Small Motors.—A description of the various small Engines useful for domestic purposes, from 1 man to 1 horse power, worked by various methods, such as Electric Engines, Gas Engines, Petroleum Engines, Steam Engines, Condensing Engines, Water Power, Wind Power, and the various methods of working and managing them.
Household Law.—The Law relating to Landlords and Tenants, Lodgers, Servants, Parochial Authorities, Juries, Insurance, Nuisance, etc.

On Designing Belt Gearing. By E. J. COWLING WELCH, Mem. Inst. Mech. Engineers, Author of 'Designing Valve Gearing.' Fcap. 8vo, sewed, 6*d.*

A Handbook of Formulæ, Tables, and Memoranda, for Architectural Surveyors and others engaged in Building. By J. T. HURST, C.E. Fourteenth edition, royal 32mo, roan, 5*s.*

"It is no disparagement to the many excellent publications we refer to, to say that in our opinion this little pocket-book of Hurst's is the very best of them all, without any exception. It would be useless to attempt a recapitulation of the contents, for it appears to contain almost *everything* that anyone connected with building could require, and, best of all, made up in a compact form for carrying in the pocket, measuring only 5 in. by 3 in., and about ⅜ in. thick, in a limp cover. We congratulate the author on the success of his laborious and practically compiled little book, which has received unqualified and deserved praise from every professional person to whom we have shown it."—*The Dublin Builder.*

Tabulated Weights of Angle, Tee, Bulb, Round, Square, and Flat Iron and Steel, and other information for the use of Naval Architects and Shipbuilders. By C. H. JORDAN, M.I.N.A. Fourth edition, 32mo, cloth, 2*s.* 6*d.*

A Complete Set of Contract Documents for a Country Lodge, comprising Drawings, Specifications, Dimensions (for quantities), Abstracts, Bill of Quantities, Form of Tender and Contract, with Notes by J. LEANING, printed in facsimile of the original documents, on single sheets fcap., in paper case, 10*s.*

A Practical Treatise on Heat, as applied to the Useful Arts; for the Use of Engineers, Architects, &c. By THOMAS BOX. With 14 *plates.* Third edition, crown 8vo, cloth, 12*s.* 6*d.*

A Descriptive Treatise on Mathematical Drawing Instruments: their construction, uses, qualities, selection, preservation, and suggestions for improvements, with hints upon Drawing and Colouring. By W. F. STANLEY, M.R.I. Fifth edition, *with numerous illustrations,* crown 8vo, cloth, 5*s.*

Quantity Surveying. By J. LEANING. With 42 illustrations. Second edition, revised, crown 8vo, cloth, 9s.

CONTENTS:

A complete Explanation of the London Practice.
General Instructions.
Order of Taking Off.
Modes of Measurement of the various Trades.
Use and Waste.
Ventilation and Warming.
Credits, with various Examples of Treatment.
Abbreviations.
Squaring the Dimensions.
Abstracting, with Examples in illustration of each Trade.
Billing.
Examples of Preambles to each Trade.
Form for a Bill of Quantities.
Do. Bill of Credits.
Do. Bill for Alternative Estimate.
Restorations and Repairs, and Form of Bill.
Variations before Acceptance of Tender.
Errors in a Builder's Estimate.
Schedule of Prices.
Form of Schedule of Prices.
Analysis of Schedule of Prices.
Adjustment of Accounts.
Form of a Bill of Variations.
Remarks on Specifications.
Prices and Valuation of Work, with Examples and Remarks upon each Trade.
The Law as it affects Quantity Surveyors, with Law Reports.
Taking Off after the Old Method.
Northern Practice.
The General Statement of the Methods recommended by the Manchester Society of Architects for taking Quantities.
Examples of Collections.
Examples of "Taking Off" in each Trade.
Remarks on the Past and Present Methods of Estimating.

Spons' Architects' and Builders' Pocket-Book of Prices and Memoranda. Edited by W. YOUNG, Architect. Crown 8vo, cloth, *Published annually*. Fifteenth edition. *Now ready*.

Long-Span Railway Bridges, comprising Investigations of the Comparative Theoretical and Practical Advantages of the various adopted or proposed Type Systems of Construction, with numerous Formulæ and Tables giving the weight of Iron or Steel required in Bridges from 300 feet to the limiting Spans; to which are added similar Investigations and Tables relating to Short-span Railway Bridges. Second and revised edition. By B. BAKER, Assoc. Inst. C.E. *Plates*, crown 8vo, cloth, 5s.

Elementary Theory and Calculation of Iron Bridges and Roofs. By AUGUST RITTER, Ph.D., Professor at the Polytechnic School at Aix-la-Chapelle. Translated from the third German edition, by H. R. SANKEY, Capt. R.E. With 500 *illustrations*, 8vo, cloth, 15s.

The Elementary Principles of Carpentry. By THOMAS TREDGOLD. Revised from the original edition, and partly re-written, by JOHN THOMAS HURST. Contained in 517 pages of letter-press, and *illustrated with 48 plates and 150 wood engravings*. Sixth edition, reprinted from the third, crown 8vo, cloth, 12s. 6d.

Section I. On the Equality and Distribution of Forces — Section II. Resistance of Timber — Section III. Construction of Floors — Section IV. Construction of Roofs — Section V. Construction of Domes and Cupolas — Section VI. Construction of Partitions — Section VII. Scaffolds, Staging, and Gantries — Section VIII. Construction of Centres for Bridges — Section IX. Coffer-dams, Shoring, and Strutting — Section X. Wooden Bridges and Viaducts — Section XI. Joints, Straps, and other Fastenings — Section XII. Timber.

The Builder's Clerk: a Guide to the Management of a Builder's Business. By THOMAS BALES. Fcap. 8vo, cloth, 1s. 6d.

Our Factories, Workshops, and Warehouses: their Sanitary and Fire-Resisting Arrangements. By B. H. THWAITE, Assoc. Mem. Inst. C.E. *With* 183 *wood engravings*, crown 8vo, cloth, 9s.

Gold: Its Occurrence and Extraction, embracing the Geographical and Geological Distribution and the Mineralogical Characters of Gold-bearing rocks; the peculiar features and modes of working Shallow Placers, Rivers, and Deep Leads; Hydraulicing; the Reduction and Separation of Auriferous Quartz; the treatment of complex Auriferous ores containing other metals; a Bibliography of the subject and a Glossary of Technical and Foreign Terms. By ALFRED G. LOCK, F.R.G.S. *With numerous illustrations and maps*, 1250 pp., super-royal 8vo, cloth, 2l. 12s. 6d.

Iron Roofs: Examples of Design, Description. *Illustrated with* 64 *Working Drawings of Executed Roofs.* By ARTHUR T. WALMISLEY, Assoc. Mem. Inst. C.E. Second edition, revised, imp. 4to, half-morocco, 3l. 3s.

A History of Electric Telegraphy, to the Year 1837. Chiefly compiled from Original Sources, and hitherto Unpublished Documents, by J. J. FAHIE, Mem. Soc. of Tel. Engineers, and of the International Society of Electricians, Paris. Crown 8vo, cloth, 9s.

Spons' Information for Colonial Engineers. Edited by J. T. HURST. Demy 8vo, sewed.

No. 1, Ceylon. By ABRAHAM DEANE, C.E. 2s. 6d.

CONTENTS:

Introductory Remarks—Natural Productions—Architecture and Engineering—Topography, Trade, and Natural History—Principal Stations—Weights and Measures, etc., etc.

No. 2. Southern Africa, including the Cape Colony, Natal, and the Dutch Republics. By HENRY HALL, F.R.G.S., F.R.C.I. With Map. 3s. 6d.

CONTENTS:

General Description of South Africa—Physical Geography with reference to Engineering Operations—Notes on Labour and Material in Cape Colony—Geological Notes on Rock Formation in South Africa—Engineering Instruments for Use in South Africa—Principal Public Works in Cape Colony: Railways, Mountain Roads and Passes, Harbour Works, Bridges, Gas Works, Irrigation and Water Supply, Lighthouses, Drainage and Sanitary Engineering, Public Buildings, Mines—Table of Woods in South Africa—Animals used for Draught Purposes—Statistical Notes—Table of Distances—Rates of Carriage, etc.

No. 3. India. By F. C. DANVERS, Assoc. Inst. C.E. With Map. 4s. 6d.

CONTENTS:

Physical Geography of India—Building Materials—Roads—Railways—Bridges—Irrigation—River Works—Harbours—Lighthouse Buildings—Native Labour—The Principal Trees of India—Money—Weights and Measures—Glossary of Indian Terms, etc.

A Practical Treatise on Coal Mining. By GEORGE G. ANDRÉ, F.G.S., Assoc. Inst. C.E., Member of the Society of Engineers. With 82 *lithographic plates.* 2 vols., royal 4to, cloth, 3*l.* 12*s.*

A Practical Treatise on Casting and Founding, including descriptions of the modern machinery employed in the art. By N. E. SPRETSON, Engineer. Third edition, with 82 *plates* drawn to scale, 412 pp., demy 8vo, cloth, 18*s.*

The Depreciation of Factories and their Valuation. By EWING MATHESON, M. Inst. C.E. 8vo, cloth, 6*s.*

A Handbook of Electrical Testing. By H. R. KEMPE, M.S.T.E. Fourth edition, revised and enlarged, crown 8vo, cloth, 16*s.*

Gas Works: their Arrangement, Construction, Plant, and Machinery. By F. COLYER, M. Inst. C.E. *With* 31 *folding plates,* 8vo, cloth, 24*s.*

The Clerk of Works: a Vade-Mecum for all engaged in the Superintendence of Building Operations. By G. G. HOSKINS, F.R.I.B.A. Third edition, fcap. 8vo, cloth, 1*s.* 6*d.*

American Foundry Practice: Treating of Loam, Dry Sand, and Green Sand Moulding, and containing a Practical Treatise upon the Management of Cupolas, and the Melting of Iron. By T. D. WEST, Practical Iron Moulder and Foundry Foreman. Second edition, *with numerous illustrations,* crown 8vo, cloth, 10*s.* 6*d.*

The Maintenance of Macadamised Roads. By T. CODRINGTON, M.I.C.E, F.G.S., General Superintendent of County Roads for South Wales. 8vo, cloth, 6*s.*

Hydraulic Steam and Hand Power Lifting and Pressing Machinery. By FREDERICK COLYER, M. Inst. C.E., M. Inst. M.E. With 73 *plates,* 8vo, cloth, 18*s.*

Pumps and Pumping Machinery. By F. COLYER, M.I.C.E., M.I.M.E. *With* 23 *folding plates,* 8vo, cloth, 12*s.* 6*d.*

Pumps and Pumping Machinery. By F. COLYER. Second Part. *With* 11 *large plates,* 8vo, cloth, 12*s.* 6*d.*

A Treatise on the Origin, Progress, Prevention, and Cure of Dry Rot in Timber; with Remarks on the Means of Preserving Wood from Destruction by Sea-Worms, Beetles, Ants, etc. By THOMAS ALLEN BRITTON, late Surveyor to the Metropolitan Board of Works, etc., etc. *With* 10 *plates,* crown 8vo, cloth, 7*s.* 6*d.*

The Municipal and Sanitary Engineer's Handbook.
By H. PERCY BOULNOIS, Mem. Inst. C.E., Borough Engineer, Portsmouth. With *numerous illustrations*, demy 8vo, cloth, 12s. 6d.

CONTENTS:

The Appointment and Duties of the Town Surveyor—Traffic—Macadamised Roadways—Steam Rolling—Road Metal and Breaking—Pitched Pavements—Asphalte—Wood Pavements—Footpaths—Kerbs and Gutters—Street Naming and Numbering—Street Lighting—Sewerage—Ventilation of Sewers—Disposal of Sewage—House Drainage—Disinfection—Gas and Water Companies, etc., Breaking up Streets—Improvement of Private Streets—Borrowing Powers—Artizans' and Labourers' Dwellings—Public Conveniences—Scavenging, including Street Cleansing—Watering and the Removing of Snow—Planting Street Trees—Deposit of Plans—Dangerous Buildings—Hoardings—Obstructions—Improving Street Lines—Cellar Openings—Public Pleasure Grounds—Cemeteries—Mortuaries—Cattle and Ordinary Markets—Public Slaughter-houses, etc.—Giving numerous Forms of Notices, Specifications, and General Information upon these and other subjects of great importance to Municipal Engineers and others engaged in Sanitary Work.

Metrical Tables. By G. L. MOLESWORTH, M.I.C.E.
32mo, cloth, 1s. 6d.

CONTENTS.

General—Linear Measures—Square Measures—Cubic Measures—Measures of Capacity—Weights—Combinations—Thermometers.

Elements of Construction for Electro-Magnets. By
Count TH. DU MONCEL, Mem. de l'Institut de France. Translated from the French by C. J. WHARTON. Crown 8vo, cloth, 4s. 6d.

Practical Electrical Units Popularly Explained, with
numerous illustrations and Remarks. By JAMES SWINBURNE, late of J. W. Swan and Co., Paris, late of Brush-Swan Electric Light Company, U.S.A. 18mo, cloth, 1s. 6d.

A Treatise on the Use of Belting for the Transmission of Power.
By J. H. COOPER. Second edition, *illustrated*, 8vo, cloth, 15s.

A Pocket-Book of Useful Formulæ and Memoranda for Civil and Mechanical Engineers.
By GUILFORD L. MOLESWORTH, Mem. Inst. C.E., Consulting Engineer to the Government of India for State Railways. *With numerous illustrations*, 744 pp. Twenty-first edition, revised and enlarged, 32mo, roan, 6s.

SYNOPSIS OF CONTENTS:

Surveying, Levelling, etc.—Strength and Weight of Materials—Earthwork, Brickwork, Masonry, Arches, etc.—Struts, Columns, Beams, and Trusses—Flooring, Roofing, and Roof Trusses—Girders, Bridges, etc.—Railways and Roads—Hydraulic Formulæ—Canals, Sewers, Waterworks, Docks—Irrigation and Breakwaters—Gas, Ventilation, and Warming—Heat, Light, Colour, and Sound—Gravity: Centres, Forces, and Powers—Millwork, Teeth of Wheels, Shafting, etc.—Workshop Recipes—Sundry Machinery—Animal Power—Steam and the Steam Engine—Water-power, Water-wheels, Turbines, etc.—Wind and Windmills—Steam Navigation, Ship Building, Tonnage, etc.—Gunnery, Projectiles, etc.—Weights, Measures, and Money—Trigonometry, Conic Sections, and Curves—Telegraphy—Mensuration—Tables of Areas and Circumference, and Arcs of Circles—Logarithms, Square and Cube Roots, Powers—Reciprocals, etc.—Useful Numbers—Differential and Integral Calculus—Algebraic Signs—Telegraphic Construction and Formulæ.

Hints on Architectural Draughtsmanship. By G. W. TUXFORD HALLATT. Fcap. 8vo, cloth, 1s. 6d.

Spons' Tables and Memoranda for Engineers; selected and arranged by J. T. HURST, C.E., Author of 'Architectural Surveyors' Handbook,' 'Hurst's Tredgold's Carpentry,' etc. Ninth edition, 64mo, roan, gilt edges, 1s.; or in cloth case, 1s. 6d.

This work is printed in a pearl type, and is so small, measuring only 2½ in. by 1¾ in. by ¼ in. thick, that it may be easily carried in the waistcoat pocket.

"It is certainly an extremely rare thing for a reviewer to be called upon to notice a volume measuring but 2½ in. by 1¾ in., yet these dimensions faithfully represent the size of the handy little book before us. The volume—which contains 118 printed pages, besides a few blank pages for memoranda—is, in fact, a true pocket-book, adapted for being carried in the waistcoat pocket, and containing a far greater amount and variety of information than most people would imagine could be compressed into so small a space. . . . The little volume has been compiled with considerable care and judgment, and we can cordially recommend it to our readers as a useful little pocket companion."—*Engineering.*

A Practical Treatise on Natural and Artificial Concrete, its Varieties and Constructive Adaptations. By HENRY REID, Author of the 'Science and Art of the Manufacture of Portland Cement.' New Edition, *with 59 woodcuts and 5 plates,* 8vo, cloth, 15s.

Notes on Concrete and Works in Concrete; especially written to assist those engaged upon Public Works. By JOHN NEWMAN, Assoc. Mem. Inst. C.E., crown 8vo, cloth, 4s. 6d.

Electricity as a Motive Power. By Count TH. DU MONCEL, Membre de l'Institut de France, and FRANK GERALDY, Ingénieur des Ponts et Chaussées. Translated and Edited, with Additions, by C. J. WHARTON, Assoc. Soc. Tel. Eng. and Elec. *With 113 engravings and diagrams,* crown 8vo, cloth, 7s. 6d.

Treatise on Valve-Gears, with special consideration of the Link-Motions of Locomotive Engines. By Dr. GUSTAV ZEUNER, Professor of Applied Mechanics at the Confederated Polytechnikum of Zurich. Translated from the Fourth German Edition, by Professor J. F. KLEIN, Lehigh University, Bethlehem, Pa. *Illustrated,* 8vo, cloth, 12s. 6d.

The French-Polisher's Manual. By a French-Polisher; containing Timber Staining, Washing, Matching, Improving, Painting, Imitations, Directions for Staining, Sizing, Embodying, Smoothing, Spirit Varnishing, French-Polishing, Directions for Re-polishing. Third edition, royal 32mo, sewed, 6d.

Hops, their Cultivation, Commerce, and Uses in various Countries. By P. L. SIMMONDS. Crown 8vo, cloth, 4s. 6d.

The Principles of Graphic Statics. By GEORGE SYDENHAM CLARKE, Capt. Royal Engineers. *With 112 illustrations.* 4to, cloth, 12s. 6d.

Dynamo-Electric Machinery: A Manual for Students
of Electro-technics. By SILVANUS P. THOMPSON, B.A., D.Sc., Professor of Experimental Physics in University College, Bristol, etc., etc. Second edition, *illustrated*, 8vo, cloth, 12s. 6d.

Practical Geometry, Perspective, and Engineering Drawing; a Course of Descriptive Geometry adapted to the Requirements of the Engineering Draughtsman, including the determination of cast shadows and Isometric Projection, each chapter being followed by numerous examples; to which are added rules for Shading, Shade-lining, etc., together with practical instructions as to the Lining, Colouring, Printing, and general treatment of Engineering Drawings, with a chapter on drawing Instruments. By GEORGE S. CLARKE, Capt. R.E. Second edition, *with 21 plates*. 2 vols., cloth, 10s. 6d.

The Elements of Graphic Statics. By Professor KARL VON OTT, translated from the German by G. S. CLARKE, Capt. R.E., Instructor in Mechanical Drawing, Royal Indian Engineering College. *With 93 illustrations*, crown 8vo, cloth, 5s.

A Practical Treatise on the Manufacture and Distribution of Coal Gas. By WILLIAM RICHARDS. Demy 4to, with *numerous wood engravings and 29 plates*, cloth, 28s.

SYNOPSIS OF CONTENTS:

Introduction — History of Gas Lighting — Chemistry of Gas Manufacture, by Lewis Thompson, Esq., M.R.C.S.—Coal, with Analyses, by J. Paterson, Lewis Thompson, and G. R. Hislop, Esqrs.—Retorts, Iron and Clay—Retort Setting—Hydraulic Main—Condensers — Exhausters — Washers and Scrubbers — Purifiers — Purification — History of Gas Holder — Tanks, Brick and Stone, Composite, Concrete, Cast-iron, Compound Annular Wrought-iron — Specifications — Gas Holders — Station Meter — Governor — Distribution— Mains—Gas Mathematics, or Formulæ for the Distribution of Gas, by Lewis Thompson, Esq.— Services—Consumers' Meters—Regulators—Burners—Fittings—Photometer—Carburization of Gas—Air Gas and Water Gas—Composition of Coal Gas, by Lewis Thompson, Esq.— Analyses of Gas—Influence of Atmospheric Pressure and Temperature on Gas—Residual Products—Appendix—Description of Retort Settings, Buildings, etc., etc.

The New Formula for Mean Velocity of Discharge of Rivers and Canals. By W. R. KUTTER. Translated from articles in the 'Cultur-Ingénieur,' by LOWIS D'A. JACKSON, Assoc. Inst. C.E. 8vo, cloth, 12s. 6d.

The Practical Millwright and Engineer's Ready Reckoner; or Tables for finding the diameter and power of cog-wheels, diameter, weight, and power of shafts, diameter and strength of bolts, etc. By THOMAS DIXON. Fourth edition, 12mo, cloth, 3s.

Tin: Describing the Chief Methods of Mining, Dressing and Smelting it abroad; with Notes upon Arsenic, Bismuth and Wolfram. By ARTHUR G. CHARLETON, Mem. American Inst. of Mining Engineers. *With plates*, 8vo, cloth, 12s. 6d.

Perspective, Explained and Illustrated. By G. S. CLARKE, Capt. R.E. *With illustrations*, 8vo, cloth, 3s. 6d.

Practical Hydraulics; a Series of Rules and Tables for the use of Engineers, etc., etc. By THOMAS BOX. Fifth edition, *numerous plates*, post 8vo, cloth, 5s.

The Essential Elements of Practical Mechanics; based on the Principle of Work, designed for Engineering Students. By OLIVER BYRNE, formerly Professor of Mathematics, College for Civil Engineers. Third edition, *with* 148 *wood engravings*, post 8vo, cloth, 7s. 6d.

CONTENTS:

Chap. 1. How Work is Measured by a Unit, both with and without reference to a Unit of Time—Chap. 2. The Work of Living Agents, the Influence of Friction, and introduces one of the most beautiful Laws of Motion—Chap. 3. The principles expounded in the first and second chapters are applied to the Motion of Bodies—Chap. 4. The Transmission of Work by simple Machines—Chap. 5. Useful Propositions and Rules.

Breweries and Maltings: their Arrangement, Construction, Machinery, and Plant. By G. SCAMELL, F.R.I.B.A. Second edition, revised, enlarged, and partly rewritten. By F. COLYER, M.I.C.E., M.I.M.E. *With* 20 *plates*, 8vo, cloth, 18s.

A Practical Treatise on the Construction of Horizontal and Vertical Waterwheels, specially designed for the use of operative mechanics. By WILLIAM CULLEN, Millwright and Engineer. *With* 11 *plates*. Second edition, revised and enlarged, small 4to, cloth, 12s. 6d.

A Practical Treatise on Mill-gearing, Wheels, Shafts, Riggers, etc.; for the use of Engineers. By THOMAS BOX. Third edition, *with* 11 *plates*. Crown 8vo, cloth, 7s. 6d.

Mining Machinery: a Descriptive Treatise on the Machinery, Tools, and other Appliances used in Mining. By G. G. ANDRÉ, F.G.S., Assoc. Inst. C.E., Mem. of the Society of Engineers. Royal 4to, uniform with the Author's Treatise on Coal Mining, containing 182 *plates*, accurately drawn to scale, with descriptive text, in 2 vols., cloth, 3l. 12s.

CONTENTS:

Machinery for Prospecting, Excavating, Hauling, and Hoisting—Ventilation—Pumping—Treatment of Mineral Products, including Gold and Silver, Copper, Tin, and Lead, Iron Coal, Sulphur, China Clay, Brick Earth, etc.

Tables for Setting out Curves for Railways, Canals, Roads, etc., varying from a radius of five chains to three miles. By A. KENNEDY and R. W. HACKWOOD. *Illustrated*, 32mo, cloth, 2s. 6d.

The Science and Art of the Manufacture of Portland Cement, with observations on some of its constructive applications. With 66 *illustrations*. By HENRY REID, C.E., Author of 'A Practical Treatise on Concrete,' etc., etc. 8vo, cloth, 18*s*.

The Draughtsman's Handbook of Plan and Map Drawing; including instructions for the preparation of Engineering, Architectural, and Mechanical Drawings. *With numerous illustrations in the text, and* 33 *plates* (15 *printed in colours*). By G. G. ANDRÉ, F.G.S., Assoc. Inst. C.E. 4to, cloth, 9*s*.

CONTENTS:

The Drawing Office and its Furnishings—Geometrical Problems—Lines, Dots, and their Combinations—Colours, Shading, Lettering, Bordering, and North Points—Scales—Plotting—Civil Engineers' and Surveyors' Plans—Map Drawing—Mechanical and Architectural Drawing—Copying and Reducing Trigonometrical Formulæ, etc., etc.

The Boiler-maker's and Iron Ship-builder's Companion, comprising a series of original and carefully calculated tables, of the utmost utility to persons interested in the iron trades. By JAMES FODEN, author of 'Mechanical Tables,' etc. Second edition revised, *with illustrations*, crown 8vo, cloth, 5*s*.

Rock Blasting: a Practical Treatise on the means employed in Blasting Rocks for Industrial Purposes. By G. G. ANDRÉ, F.G.S., Assoc. Inst. C.E. *With* 56 *illustrations and* 12 *plates*, 8vo, cloth, 10*s*. 6*d*.

Painting and Painters' Manual: a Book of Facts for Painters and those who Use or Deal in Paint Materials. By C. L. CONDIT and J. SCHELLER. *Illustrated*, 8vo, cloth, 10*s*. 6*d*.

A Treatise on Ropemaking as practised in public and private Rope-yards, with a Description of the Manufacture, Rules, Tables of Weights, etc., adapted to the Trade, Shipping, Mining, Railways, Builders, etc. By R. CHAPMAN, formerly foreman to Messrs. Huddart and Co., Limehouse, and late Master Ropemaker to H.M. Dockyard, Deptford. Second edition, 12mo, cloth, 3*s*.

Laxton's Builders' and Contractors' Tables; for the use of Engineers, Architects, Surveyors, Builders, Land Agents, and others. Bricklayer, containing 22 tables, with nearly 30,000 calculations. 4to, cloth, 5*s*.

Laxton's Builders' and Contractors' Tables. Excavator, Earth, Land, Water, and Gas, containing 53 tables, with nearly 24,000 calculations. 4to, cloth, 5*s*.

Sanitary Engineering: a Guide to the Construction of Works of Sewerage and House Drainage, with Tables for facilitating the calculations of the Engineer. By BALDWIN LATHAM, C.E., M. Inst. C.E., F.G.S., F.M.S., Past-President of the Society of Engineers. Second edition, *with numerous plates and woodcuts*, 8vo, cloth, 1*l*. 10*s*.

Screw Cutting Tables for Engineers and Machinists, giving the values of the different trains of Wheels required to produce Screws of any pitch, calculated by Lord Lindsay, M.P., F.R.S., F.R.A.S., etc. Cloth, oblong, 2*s*.

Screw Cutting Tables, for the use of Mechanical Engineers, showing the proper arrangement of Wheels for cutting the Threads of Screws of any required pitch, with a Table for making the Universal Gas-pipe Threads and Taps. By W. A. MARTIN, Engineer. Second edition, oblong, cloth, 1*s*., or sewed, 6*d*.

A Treatise on a Practical Method of Designing Slide-Valve Gears by Simple Geometrical Construction, based upon the principles enunciated in Euclid's Elements, and comprising the various forms of Plain Slide-Valve and Expansion Gearing; together with Stephenson's, Gooch's, and Allan's Link-Motions, as applied either to reversing or to variable expansion combinations. By EDWARD J. COWLING WELCH, Memb. Inst. Mechanical Engineers. Crown 8vo, cloth, 6*s*.

Cleaning and Scouring: a Manual for Dyers, Laundresses, and for Domestic Use. By S. CHRISTOPHER. 18mo, sewed, 6*d*.

A Glossary of Terms used in Coal Mining. By WILLIAM STUKELEY GRESLEY, Assoc. Mem. Inst. C.E., F.G.S., Member of the North of England Institute of Mining Engineers. *Illustrated with numerous woodcuts and diagrams*, crown 8vo, cloth, 5*s*.

A Pocket-Book for Boiler Makers and Steam Users, comprising a variety of useful information for Employer and Workman, Government Inspectors, Board of Trade Surveyors, Engineers in charge of Works and Slips, Foremen of Manufactories, and the general Steam-using Public. By MAURICE JOHN SEXTON. Second edition, royal 32mo, roan, gilt edges, 5*s*.

Electrolysis: a Practical Treatise on Nickeling, Coppering, Gilding, Silvering, the Refining of Metals, and the treatment of Ores by means of Electricity. By HIPPOLYTE FONTAINE, translated from the French by J. A. BERLY, C.E., Assoc. S.T.E. *With engravings.* 8vo, cloth, 9*s*.

PUBLISHED BY E. & F. N. SPON. 13

Barlow's Tables of Squares, Cubes, Square Roots,
Cube Roots, Reciprocals of all Integer Numbers up to 10,000. Post 8vo, cloth, 6s.

A Practical Treatise on the Steam Engine, containing Plans and Arrangements of Details for Fixed Steam Engines, with Essays on the Principles involved in Design and Construction. By ARTHUR RIGG, Engineer, Member of the Society of Engineers and of the Royal Institution of Great Britain. Demy 4to, *copiously illustrated with woodcuts and* 96 *plates,* in one Volume, half-bound morocco, 2l. 2s.; or cheaper edition, cloth, 25s.

This work is not, in any sense, an elementary treatise, or history of the steam engine, but is intended to describe examples of Fixed Steam Engines without entering into the wide domain of locomotive or marine practice. To this end illustrations will be given of the most recent arrangements of Horizontal, Vertical, Beam, Pumping, Winding, Portable, Semi-portable, Corliss, Allen, Compound, and other similar Engines, by the most eminent Firms in Great Britain and America. The laws relating to the action and precautions to be observed in the construction of the various details, such as Cylinders, Pistons, Piston-rods, Connecting-rods, Cross-heads, Motion-blocks, Eccentrics, Simple, Expansion, Balanced, and Equilibrium Slide-valves, and Valve-gearing will be minutely dealt with. In this connection will be found articles upon the Velocity of Reciprocating Parts and the Mode of Applying the Indicator, Heat and Expansion of Steam Governors, and the like. It is the writer's desire to draw illustrations from every possible source, and give only those rules that present practice deems correct.

A Practical Treatise on the Science of Land and Engineering Surveying, Levelling, Estimating Quantities, etc., with a general description of the several Instruments required for Surveying, Levelling, Plotting, etc. By H. S. MERRETT. Fourth edition, revised by G. W. USILL, Assoc. Mem. Inst. C.E. 41 *plates, with illustrations and tables,* royal 8vo, cloth, 12s. 6d.

PRINCIPAL CONTENTS:

Part 1. Introduction and the Principles of Geometry. Part 2. Land Surveying; comprising General Observations—The Chain—Offsets Surveying by the Chain only—Surveying Hilly Ground—To Survey an Estate or Parish by the Chain only—Surveying with the Theodolite—Mining and Town Surveying—Railroad Surveying—Mapping—Division and Laying out of Land—Observations on Enclosures—Plane Trigonometry. Part 3. Levelling—Simple and Compound Levelling—The Level Book—Parliamentary Plan and Section—Levelling with a Theodolite—Gradients—Wooden Curves—To Lay out a Railway Curve—Setting out Widths. Part 4. Calculating Quantities generally for Estimates—Cuttings and Embankments—Tunnels—Brickwork—Ironwork—Timber Measuring. Part 5. Description and Use of Instruments in Surveying and Plotting—The Improved Dumpy Level—Troughton's Level—The Prismatic Compass—Proportional Compass—Box Sextant—Vernier—Pantagraph—Merrett's Improved Quadrant—Improved Computation Scale—The Diagonal Scale—Straight Edge and Sector. Part 6. Logarithms of Numbers—Logarithmic Sines and Co-Sines, Tangents and Co-Tangents—Natural Sines and Co-Sines—Tables for Earthwork, for Setting out Curves, and for various Calculations, etc., etc., etc.

Health and Comfort in House Building, or Ventilation with Warm Air by Self-Acting Suction Power, with Review of the mode of Calculating the Draught in Hot-Air Flues, and with some actual Experiments. By J. DRYSDALE, M.D., and J. W. HAYWARD, M.D. Second edition, with Supplement, *with plates,* demy 8vo, cloth, 7s. 6d.

The Assayer's Manual: an Abridged Treatise on the Docimastic Examination of Ores and Furnace and other Artificial Products. By BRUNO KERL. Translated by W. T. BRANNT. *With 65 illustrations*, 8vo, cloth, 12s. 6d.

Electricity: its Theory, Sources, and Applications. By J. T. SPRAGUE, M.S.T.E. Second edition, revised and enlarged, *with numerous illustrations*, crown 8vo, cloth, 15s.

The Practice of Hand Turning in Wood, Ivory, Shell, etc., with Instructions for Turning such Work in Metal as may be required in the Practice of Turning in Wood, Ivory, etc.; also an Appendix on Ornamental Turning. (A book for beginners.) By FRANCIS CAMPIN. Third edition, *with wood engravings*, crown 8vo, cloth, 6s.

CONTENTS:

On Lathes—Turning Tools—Turning Wood—Drilling—Screw Cutting—Miscellaneous Apparatus and Processes—Turning Particular Forms—Staining—Polishing—Spinning Metals—Materials—Ornamental Turning, etc.

Treatise on Watchwork, Past and Present. By the Rev. H. L. NELTHROPP, M.A., F.S.A. *With 32 illustrations*, crown 8vo, cloth, 6s. 6d.

CONTENTS:

Definitions of Words and Terms used in Watchwork—Tools—Time—Historical Summary—On Calculations of the Numbers for Wheels and Pinions; their Proportional Sizes, Trains, etc.—Of Dial Wheels, or Motion Work—Length of Time of Going without Winding up—The Verge—The Horizontal—The Duplex—The Lever—The Chronometer—Repeating Watches—Keyless Watches—The Pendulum, or Spiral Spring—Compensation—Jewelling of Pivot Holes—Clerkenwell—Fallacies of the Trade—Incapacity of Workmen—How to Choose and Use a Watch, etc.

Algebra Self-Taught. By W. P. HIGGS, M.A., D.Sc., LL.D., Assoc. Inst. C.E., Author of 'A Handbook of the Differential Calculus,' etc. Second edition, crown 8vo, cloth, 2s. 6d.

CONTENTS:

Symbols and the Signs of Operation—The Equation and the Unknown Quantity—Positive and Negative Quantities—Multiplication—Involution—Exponents—Negative Exponents—Roots, and the Use of Exponents as Logarithms—Logarithms—Tables of Logarithms and Proportionate Parts—Transformation of System of Logarithms—Common Uses of Common Logarithms—Compound Multiplication and the Binomial Theorem—Division, Fractions, and Ratio—Continued Proportion—The Series and the Summation of the Series—Limit of Series—Square and Cube Roots—Equations—List of Formulæ, etc.

Spons' Dictionary of Engineering, Civil, Mechanical, Military, and Naval; with technical terms in French, German, Italian, and Spanish, 3100 pp., and *nearly 8000 engravings*, in super-royal 8vo, in 8 divisions, 5l. 8s. Complete in 3 vols., cloth, 5l. 5s. Bound in a superior manner, half-morocco, top edge gilt, 3 vols., 6l. 12s.

Notes in Mechanical Engineering. Compiled principally for the use of the Students attending the Classes on this subject at the City of London College. By HENRY ADAMS, Mem. Inst. M.E., Mem. Inst. C.E., Mem. Soc. of Engineers. Crown 8vo, cloth, 2s. 6d.

Canoe and Boat Building: a complete Manual for Amateurs, containing plain and comprehensive directions for the construction of Canoes, Rowing and Sailing Boats, and Hunting Craft. By W. P. STEPHENS. *With numerous illustrations and 24 plates of Working Drawings.* Crown 8vo, cloth, 7s. 6d.

Proceedings of the National Conference of Electricians, Philadelphia, October 8th to 13th, 1884. 18mo, cloth, 3s.

Dynamo - Electricity, its Generation, Application, Transmission, Storage, and Measurement. By G. B. PRESCOTT. With 545 *illustrations.* 8vo, cloth, 1l. 1s.

Domestic Electricity for Amateurs. Translated from the French of E. HOSPITALIER, Editor of "L'Electricien," by C. J. WHARTON, Assoc. Soc. Tel. Eng. *Numerous illustrations.* Demy 8vo, cloth, 9s.

CONTENTS:

1. Production of the Electric Current—2. Electric Bells—3. Automatic Alarms—4. Domestic Telephones—5. Electric Clocks—6. Electric Lighters—7. Domestic Electric Lighting—8. Domestic Application of the Electric Light—9. Electric Motors—10. Electrical Locomotion—11. Electrotyping, Plating, and Gilding—12. Electric Recreations—13. Various applications—Workshop of the Electrician.

Wrinkles in Electric Lighting. By VINCENT STEPHEN. *With illustrations.* 18mo, cloth, 2s. 6d.

CONTENTS:

1. The Electric Current and its production by Chemical means—2. Production of Electric Currents by Mechanical means—3. Dynamo-Electric Machines—4. Electric Lamps—5. Lead—6. Ship Lighting.

The Practical Flax Spinner; being a Description of the Growth, Manipulation, and Spinning of Flax and Tow. By LESLIE C. MARSHALL, of Belfast. *With illustrations.* 8vo, cloth, 15s.

Foundations and Foundation Walls for all classes of Buildings, Pile Driving, Building Stones and Bricks, Pier and Wall construction, Mortars, Limes, Cements, Concretes, Stuccos, &c. 64 *illustrations.* By G. T. POWELL and F. BAUMAN. 8vo, cloth, 10s. 6d.

Manual for Gas Engineering Students. By D. LEE.
18mo, cloth 1s.

Hydraulic Machinery, Past and Present. A Lecture delivered to the London and Suburban Railway Officials' Association. By H. ADAMS, Mem. Inst. C.E. *Folding plate.* 8vo, sewed, 1s.

Twenty Years with the Indicator. By THOMAS PRAY, Jun., C.E., M.E., Member of the American Society of Civil Engineers. 2 vols., royal 8vo, cloth, 12s. 6d.

Annual Statistical Report of the Secretary to the Members of the Iron and Steel Association on the Home and Foreign Iron and Steel Industries in 1884. Issued March 1885. 8vo, sewed, 5s.

Bad Drains, and How to Test them; with Notes on the Ventilation of Sewers, Drains, and Sanitary Fittings, and the Origin and Transmission of Zymotic Disease. By R. HARRIS REEVES. Crown 8vo, cloth, 3s. 6d.

Standard Practical Plumbing; being a complete Encyclopædia for Practical Plumbers and Guide for Architects, Builders, Gas Fitters, Hot-water Fitters, Ironmongers, Lead Burners, Sanitary Engineers, Zinc Workers, &c. *Illustrated by over 2000 engravings.* By P. J. DAVIES. Vol. 1, royal 8vo, cloth, 7s. 6d.

Pneumatic Transmission of Messages and Parcels between Paris and London, viâ Calais and Dover. By J. B. BERLIER, C.E. Small folio, sewed, 6d.

List of Tests (Reagents), arranged in alphabetical order, according to the names of the originators. Designed especially for the convenient reference of Chemists, Pharmacists, and Scientists. By HANS M. WILDER. Crown 8vo, cloth, 4s. 6d.

Ten Years' Experience in Works of Intermittent Downward Filtration. By J. BAILEY DENTON, Mem. Inst. C.E. Second edition, with additions. Royal 8vo, sewed, 4s.

A Treatise on the Manufacture of Soap and Candles, Lubricants and Glycerin. By W. LANT CARPENTER, B.A., B.Sc. (late of Messrs. C. Thomas and Brothers, Bristol). *With illustrations.* Crown 8vo, cloth, 10s. 6d.

The Stability of Ships explained simply, and calculated by a new Graphic method. By J. C. SPENCE, M.I.N.A. 4to, sewed, 3s. 6d.

Steam Making, or Boiler Practice. By CHARLES A. SMITH, C.E. 8vo, cloth, 10s. 6d.

CONTENTS:

1. The Nature of Heat and the Properties of Steam—2. Combustion.—3. Externally Fired Stationary Boilers—4. Internally Fired Stationary Boilers—5. Internally Fired Portable Locomotive and Marine Boilers—6. Design, Construction, and Strength of Boilers—7. Proportions of Heating Surface, Economic Evaporation, Explosions—8. Miscellaneous Boilers, Choice of Boiler Fittings and Appurtenances.

The Fireman's Guide; a Handbook on the Care of Boilers. By TEKNOLOG, föreningen T. I. Stockholm. Translated from the third edition, and revised by KARL P. DAHLSTROM, M.E. Second edition. Fcap. 8vo, cloth, 2s.

A Treatise on Modern Steam Engines and Boilers, including Land Locomotive, and Marine Engines and Boilers, for the use of Students. By FREDERICK COLYER, M. Inst. C.E., Mem. Inst. M.E. *With* 36 *plates.* 4to, cloth, 25s.

CONTENTS:

1. Introduction—2. Original Engines—3. Boilers—4. High-Pressure Beam Engines—5. Cornish Beam Engines—6. Horizontal Engines—7. Oscillating Engines—8. Vertical High-Pressure Engines—9. Special Engines—10. Portable Engines—11. Locomotive Engines—12. Marine Engines.

Steam Engine Management; a Treatise on the Working and Management of Steam Boilers. By F. COLYER, M. Inst. C.E., Mem. Inst. M.E. 18mo, cloth, 2s.

Land Surveying on the Meridian and Perpendicular System. By WILLIAM PENMAN, C.E. 8vo, cloth, 8s. 6d.

The Topographer, his Instruments and Methods, designed for the use of Students, Amateur Photographers, Surveyors, Engineers, and all persons interested in the location and construction of works based upon Topography. *Illustrated with numerous plates, maps, and engravings.* By LEWIS M. HAUPT, A.M. 8vo, cloth, 18s.

A Text-Book of Tanning, embracing the Preparation of all kinds of Leather. By HARRY R. PROCTOR, F.C.S., of Low Lights Tanneries. *With illustrations.* Crown 8vo, cloth, 10s. 6d.

In super-royal 8vo, 1168 pp., *with* 2400 *illustrations*, in 3 Divisions, cloth, price 13s. 6d. each ; or 1 vol., cloth, 2l. ; or half-morocco, 2l. 8s.

A SUPPLEMENT

TO

SPONS' DICTIONARY OF ENGINEERING.

EDITED BY ERNEST SPON, MEMB. SOC. ENGINEERS.

Abacus, Counters, Speed Indicators, and Slide Rule.
Agricultural Implements and Machinery.
Air Compressors.
Animal Charcoal Machinery.
Antimony.
Axles and Axle-boxes.
Barn Machinery.
Belts and Belting.
Blasting. Boilers.
Brakes.
Brick Machinery.
Bridges.
Cages for Mines.
Calculus, Differential and Integral.
Canals.
Carpentry.
Cast Iron.
Cement, Concrete, Limes, and Mortar.
Chimney Shafts.
Coal Cleansing and Washing.
Coal Mining.
Coal Cutting Machines.
Coke Ovens. Copper.
Docks. Drainage.
Dredging Machinery.
Dynamo - Electric and Magneto-Electric Machines.
Dynamometers.
Electrical Engineering, Telegraphy, Electric Lighting and its practical details, Telephones
Engines, Varieties of.
Explosives. Fans.
Founding, Moulding and the practical work of the Foundry.
Gas, Manufacture of.
Hammers, Steam and other Power.
Heat. Horse Power.
Hydraulics.
Hydro-geology.
Indicators. Iron.
Lifts, Hoists, and Elevators.
Lighthouses, Buoys, and Beacons.
Machine Tools.
Materials of Construction.
Meters.
Ores, Machinery and Processes employed to Dress.
Piers.
Pile Driving.
Pneumatic Transmission.
Pumps.
Pyrometers.
Road Locomotives.
Rock Drills.
Rolling Stock.
Sanitary Engineering.
Shafting.
Steel.
Steam Navvy.
Stone Machinery.
Tramways.
Well Sinking.

London: E. & F. N. SPON, 125, Strand.
New York: 35, Murray Street.

NOW COMPLETE.

With nearly 1500 *illustrations*, in super-royal 8vo, in 5 Divisions, cloth. Divisions 1 to 4, 13*s.* 6*d.* each ; Division 5, 17*s.* 6*d.* ; or 2 vols., cloth, £3 10*s.*

SPONS' ENCYCLOPÆDIA

OF THE

INDUSTRIAL ARTS, MANUFACTURES, AND COMMERCIAL PRODUCTS.

EDITED BY C. G. WARNFORD LOCK, F.L.S.

Among the more important of the subjects treated of, are the following :—

Acids, 207 pp. 220 figs.
Alcohol, 23 pp. 16 figs.
Alcoholic Liquors, 13 pp.
Alkalies, 89 pp. 78 figs.
Alloys. Alum.
Asphalt. Assaying.
Beverages, 89 pp. 29 figs.
Blacks.
Bleaching Powder, 15 pp.
Bleaching, 51 pp. 48 figs.
Candles, 18 pp. 9 figs.
Carbon Bisulphide.
Celluloid, 9 pp.
Cements. Clay.
Coal-tar Products, 44 pp. 14 figs.
Cocoa, 8 pp.
Coffee, 32 pp. 13 figs.
Cork, 8 pp. 17 figs.
Cotton Manufactures, 62 pp. 57 figs.
Drugs, 38 pp.
Dyeing and Calico Printing, 28 pp. 9 figs.
Dyestuffs, 16 pp.
Electro-Metallurgy, 13 pp.
Explosives, 22 pp. 33 figs.
Feathers.
Fibrous Substances, 92 pp. 79 figs.
Floor-cloth, 16 pp. 21 figs.
Food Preservation, 8 pp.
Fruit, 8 pp.

Fur, 5 pp.
Gas, Coal, 8 pp.
Gems.
Glass, 45 pp. 77 figs.
Graphite, 7 pp.
Hair, 7 pp.
Hair Manufactures.
Hats, 26 pp. 26 figs.
Honey. Hops.
Horn.
Ice, 10 pp. 14 figs.
Indiarubber Manufactures, 23 pp. 17 figs.
Ink, 17 pp.
Ivory.
Jute Manufactures, 11 pp., 11 figs.
Knitted Fabrics — Hosiery, 15 pp. 13 figs.
Lace, 13 pp. 9 figs.
Leather, 28 pp. 31 figs.
Linen Manufactures, 16 pp. 6 figs.
Manures, 21 pp. 30 figs.
Matches, 17 pp. 38 figs.
Mordants, 13 pp.
Narcotics, 47 pp.
Nuts, 10 pp.
Oils and Fatty Substances, 125 pp.
Paint.
Paper, 26 pp. 23 figs.
Paraffin, 8 pp. 6 figs.
Pearl and Coral, 8 pp.
Perfumes, 10 pp.

Photography, 13 pp. 20 figs.
Pigments, 9 pp. 6 figs.
Pottery, 46 pp. 57 figs.
Printing and Engraving, 20 pp. 8 figs.
Rags.
Resinous and Gummy Substances, 75 pp. 16 figs.
Rope, 16 pp. 17 figs.
Salt, 31 pp. 23 figs.
Silk, 8 pp.
Silk Manufactures, 9 pp. 11 figs.
Skins, 5 pp.
Small Wares, 4 pp.
Soap and Glycerine, 39 pp. 45 figs.
Spices, 16 pp.
Sponge, 5 pp.
Starch, 9 pp. 10 figs.
Sugar, 155 pp. 134 figs.
Sulphur.
Tannin, 18 pp.
Tea, 12 pp.
Timber, 13 pp.
Varnish, 15 pp.
Vinegar, 5 pp.
Wax, 5 pp.
Wool, 2 pp.
Woollen Manufactures, 58 pp. 39 figs.

London: E. & F. N. SPON, 125, Strand.
New York: 35, Murray Street.

Crown 8vo, cloth, with illustrations, 5s.

WORKSHOP RECEIPTS,

FIRST SERIES.

BY ERNEST SPON.

SYNOPSIS OF CONTENTS.

Bookbinding.
Bronzes and Bronzing.
Candles.
Cement.
Cleaning.
Colourwashing.
Concretes.
Dipping Acids.
Drawing Office Details.
Drying Oils.
Dynamite.
Electro - Metallurgy — (Cleaning, Dipping, Scratch-brushing, Batteries, Baths, and Deposits of every description).
Enamels.
Engraving on Wood, Copper, Gold, Silver, Steel, and Stone.
Etching and Aqua Tint.
Firework Making — (Rockets, Stars, Rains, Gerbes, Jets, Tourbillons, Candles, Fires, Lances, Lights, Wheels, Fire-balloons, and minor Fireworks).
Fluxes.
Foundry Mixtures.

Freezing.
Fulminates.
Furniture Creams, Oils, Polishes, Lacquers, and Pastes.
Gilding.
Glass Cutting, Cleaning, Frosting, Drilling, Darkening, Bending, Staining, and Painting.
Glass Making.
Glues.
Gold.
Graining.
Gums.
Gun Cotton.
Gunpowder.
Horn Working.
Indiarubber.
Japans, Japanning, and kindred processes.
Lacquers.
Lathing.
Lubricants.
Marble Working.
Matches.
Mortars.
Nitro-Glycerine.
Oils.

Paper.
Paper Hanging.
Painting in Oils, in Water Colours, as well as Fresco, House, Transparency, Sign, and Carriage Painting.
Photography.
Plastering.
Polishes.
Pottery—(Clays, Bodies, Glazes, Colours, Oils, Stains, Fluxes, Enamels, and Lustres).
Scouring.
Silvering.
Soap.
Solders.
Tanning.
Taxidermy.
Tempering Metals.
Treating Horn, Mother-o'-Pearl, and like substances.
Varnishes, Manufacture and Use of.
Veneering.
Washing.
Waterproofing.
Welding.

Besides Receipts relating to the lesser Technological matters and processes, such as the manufacture and use of Stencil Plates, Blacking, Crayons, Paste, Putty, Wax, Size, Alloys, Catgut, Tunbridge Ware, Picture Frame and Architectural Mouldings, Compos, Cameos, and others too numerous to mention.

London: E. & F. N. SPON, 125, Strand.
New York: 35, Murray Street.

Crown 8vo, cloth, 485 pages, with illustrations, 5s.

WORKSHOP RECEIPTS,
SECOND SERIES.
BY ROBERT HALDANE.

SYNOPSIS OF CONTENTS.

Acidimetry and Alkalimetry.
Albumen.
Alcohol.
Alkaloids.
Baking-powders.
Bitters.
Bleaching.
Boiler Incrustations.
Cements and Lutes.
Cleansing.
Confectionery.
Copying.

Disinfectants.
Dyeing, Staining, and Colouring.
Essences.
Extracts.
Fireproofing.
Gelatine, Glue, and Size.
Glycerine.
Gut.
Hydrogen peroxide.
Ink.
Iodine.
Iodoform.

Isinglass.
Ivory substitutes.
Leather.
Luminous bodies.
Magnesia.
Matches.
Paper.
Parchment.
Perchloric acid.
Potassium oxalate.
Preserving.

Pigments, Paint, and Painting: embracing the preparation of *Pigments*, including alumina lakes, blacks (animal, bone, Frankfort, ivory, lamp, sight, soot), blues (antimony, Antwerp, cobalt, cæruleum, Egyptian, manganate, Paris, Péligot, Prussian, smalt, ultramarine), browns (bistre, hinau, sepia, sienna, umber, Vandyke), greens (baryta, Brighton, Brunswick, chrome, cobalt, Douglas, emerald, manganese, mitis, mountain, Prussian, sap, Scheele's, Schweinfurth, titanium, verdigris, zinc), reds (Brazilwood lake, carminated lake, carmine, Cassius purple, cobalt pink, cochineal lake, colcothar, Indian red, madder lake, red chalk, red lead, vermilion), whites (alum, baryta, Chinese, lead sulphate, white lead—by American, Dutch, French, German, Kremnitz, and Pattinson processes, precautions in making, and composition of commercial samples—whiting, Wilkinson's white, zinc white), yellows (chrome, gamboge, Naples, orpiment, realgar, yellow lakes); *Paint* (vehicles, testing oils, driers, grinding, storing, applying, priming, drying, filling, coats, brushes, surface, water-colours, removing smell, discoloration; miscellaneous paints—cement paint for carton-pierre, copper paint, gold paint, iron paint, lime paints, silicated paints, steatite paint, transparent paints, tungsten paints, window paint, zinc paints); *Painting* (general instructions, proportions of ingredients, measuring paint work; carriage painting—priming paint, best putty, finishing colour, cause of cracking, mixing the paints, oils, driers, and colours, varnishing, importance of washing vehicles, re-varnishing, how to dry paint; woodwork painting).

London: E. & F. N. SPON, 125, Strand.
New York: 35, Murray Street.

JUST PUBLISHED.

Crown 8vo, cloth, 480 pages, with 183 illustrations, 5s.

WORKSHOP RECEIPTS,
THIRD SERIES.

By C. G. WARNFORD LOCK.

Uniform with the First and Second Series.

SYNOPSIS OF CONTENTS.

Alloys.	Indium.	Rubidium.
Aluminium.	Iridium.	Ruthenium.
Antimony.	Iron and Steel.	Selenium.
Barium.	Lacquers and Lacquering.	Silver.
Beryllium.	Lanthanum.	Slag.
Bismuth.	Lead.	Sodium.
Cadmium.	Lithium.	Strontium.
Cæsium.	Lubricants.	Tantalum.
Calcium.	Magnesium.	Terbium.
Cerium.	Manganese.	Thallium.
Chromium.	Mercury.	Thorium.
Cobalt.	Mica.	Tin.
Copper.	Molybdenum.	Titanium.
Didymium.	Nickel.	Tungsten.
Electrics.	Niobium.	Uranium.
Enamels and Glazes.	Osmium.	Vanadium.
Erbium.	Palladium.	Yttrium.
Gallium.	Platinum.	Zinc.
Glass.	Potassium.	Zirconium.
Gold.	Rhodium.	

London: E. & F. N. SPON, 125, Strand.
New York: 35, Murray Street.

WORKSHOP RECEIPTS,
FOURTH SERIES,
DEVOTED MAINLY TO HANDICRAFTS & MECHANICAL SUBJECTS.
By C. G. WARNFORD LOCK.

250 Illustrations, with Complete Index, and a General Index to the Four Series, 5s.

Waterproofing — rubber goods, cuprammonium processes, miscellaneous preparations.

Packing and Storing articles of delicate odour or colour, of a deliquescent character, liable to ignition, apt to suffer from insects or damp, or easily broken.

Embalming and Preserving anatomical specimens.

Leather Polishes.

Cooling Air and Water, producing low temperatures, making ice, cooling syrups and solutions, and separating salts from liquors by refrigeration.

Pumps and Siphons, embracing every useful contrivance for raising and supplying water on a moderate scale, and moving corrosive, tenacious, and other liquids.

Desiccating—air- and water-ovens, and other appliances for drying natural and artificial products.

Distilling—water, tinctures, extracts, pharmaceutical preparations, essences, perfumes, and alcoholic liquids.

Emulsifying as required by pharmacists and photographers.

Evaporating—saline and other solutions, and liquids demanding special precautions.

Filtering—water, and solutions of various kinds.

Percolating and Macerating.

Electrotyping.

Stereotyping by both plaster and paper processes.

Bookbinding in all its details.

Straw Plaiting and the fabrication of baskets, matting, etc.

Musical Instruments—the preservation, tuning, and repair of pianos, harmoniums, musical boxes, etc.

Clock and Watch Mending—adapted for intelligent amateurs.

Photography—recent development in rapid processes, handy apparatus, numerous recipes for sensitizing and developing solutions, and applications to modern illustrative purposes.

London: E. & F. N. SPON, 125, Strand.
New York: 35, Murray Street.

JUST PUBLISHED.

In demy 8vo, cloth, 600 pages, and 1420 Illustrations, 6s.

SPONS'
MECHANICS' OWN BOOK;
A MANUAL FOR HANDICRAFTSMEN AND AMATEURS.

CONTENTS.

Mechanical Drawing—Casting and Founding in Iron, Brass, Bronze, and other Alloys—Forging and Finishing Iron—Sheetmetal Working—Soldering, Brazing, and Burning—Carpentry and Joinery, embracing descriptions of some 400 Woods, over 200 Illustrations of Tools and their uses, Explanations (with Diagrams) of 116 joints and hinges, and Details of Construction of Workshop appliances, rough furniture, Garden and Yard Erections, and House Building—Cabinet-Making and Veneering—Carving and Fretcutting—Upholstery—Painting, Graining, and Marbling—Staining Furniture, Woods, Floors, and Fittings—Gilding, dead and bright, on various grounds—Polishing Marble, Metals, and Wood—Varnishing—Mechanical movements, illustrating contrivances for transmitting motion—Turning in Wood and Metals—Masonry, embracing Stonework, Brickwork, Terracotta, and Concrete—Roofing with Thatch, Tiles, Slates, Felt, Zinc, &c.—Glazing with and without putty, and lead glazing—Plastering and Whitewashing—Paper-hanging—Gas-fitting—Bell-hanging, ordinary and electric Systems—Lighting—Warming—Ventilating—Roads, Pavements, and Bridges—Hedges, Ditches, and Drains—Water Supply and Sanitation—Hints on House Construction suited to new countries.

London: E. & F. N. SPON, 125, Strand.
New York: 35, Murray Street.

www.ingramcontent.com/pod-product-compliance
Lightning Source LLC
Chambersburg PA
CBHW020227090426
42735CB00010B/1617